OPERATIONS MANAGEMENT

FOR POLYTECHNIC STUDENTS

A.P. Verma

Syllabus

OPERATIONS MANAGEMENT

State Board of Technical Education, Delhi

1. Overview of Operations Management **10%**

Concept of the term Operations. Types of Business Operations–Manufacturing, Service and Trading. Types of manufacturing Systems–Mass, Batch and Produce to Order Manufacturing Systems. Necessity and Functions of Production Planning and Control. Concept of Product Life Cycle.

2. Plant Location and Layout. Material Handling. **15%**

Plant Location decision and its importance. General factors affecting Plant Location decision. Concept of Plant Layout, factors affecting Plant Layout. Types of Plant Layouts–Product, Process, Combination and Fixed Position Layouts. Concept of Line Balancing. Introduction and Functions of Material Handling. Different material Handling equipments and their selection.

3. Forecasting **20%**

Importance of Forecasting. Types of Forecasting–Long, Medium and Short term Forecasting. Concept of Time Series in Forecasting. Techniques in Forecasting—

(i) Qualitative Techniques–Educated Guess, Executive Committee Consensus, Delphi Method, Sales Force Survey, Customer Survey, Historical Analogy and Market Research.
(ii) Quantitative Techniques–Moving Average, Weighted Moving Average, Exponential Smoothing and Regression Analysis.
(iii) Errors Estimates in Forecasting–Mean Absolute Deviation and Standard Error of Estimates.

4. Materials Management **10%**

(i) Introduction to Inventory Management–Meaning and Objectives (No numerical)
(ii) Economic order Quantity–Introduction and Graphical Representation
(iii) ABC Analysis
(iv) Purchasing Procedure–Objectives of Purchasing, Functions of purchasing deparment, Steps in Purchasing.
(v) Introduction to Modern Materials Management Techniques–JIT, KANBAN, Traceability, LIFO, FIFO.
(vi) Material Resource Planning–(MRP), ERP.

5. Project Management **10%**

(i) Meaning of the term project.
(ii) Importance of Project Management
(iii) Drawing of simple project networks (Activity on Arrow and Activity on Node)
(iv) Introduction to CPM and PERT–Critical Time Calculations.

6. Supply Chain Management **10%**

(i) Meaning of a Supply Chain
(ii) Overview of Supply Chain Management
(iii) Role of Logistics in Supply Chain Management
(iv) Make/Buy Decisions
(v) Introduction to E Commerce
(vi) Importance of Information Technology in Supply Chain Management

7. Production Scheduling **15%**

(i) Overview of Production Scheduling
(ii) Loading–Concept and Types of Loading
(iii) Types of Production Schedules–Master Production Schedule
(iv) Techniques of Scheduling–Priority Rules in Sequencing, Sequencing in jobs in one, two and three machines (Johnson's Algorithm)

8. Maintenance Management **10%**

(i) Meaning and Importance of Maintenance in an industry.
(ii) Types of Maintenance Systems
(iii) Objectives of Maintenance Management–The Bathtub Curve
(iv) Activities involved in Maintenance Management

OPERATIONS MANAGEMENT

By :

A.P. VERMA

M.Sc. Engineering, IIT (BHU), MIE (India)
Academic Counsellor in IGNOU
Formerly Engineer at TATA Engg. & Locomotive Company Ltd. (TELCO) **Jamshedpur**
Faculty Member–IMT (**Ghaziabad**)
Guest Faculty Member of National Institute of Technology (NIT) **Patna**
Guest Faculty Member of Maulana Azad College of Engg. & Tech., **Neoragarh**
Guest Faculty Member of Indian Institute of Aeronautics, **New Delhi**
Guest Faculty Member of Central Institute of Plastics Engg. and Tech., **Hajipur**
Guest Faculty Member of Global Entrepreneurship and Management Academy, **(GEMA) Patna**
Guest Faculty Member of Usha Martin Academy, **Patna**

S.K. KATARIA & SONS®

Publisher of Engineering Books

4885/109, Prakash Mahal, Dr. Subhash Bhargav Lane,
Opposite Delhi Medical Association, Daryaganj, New Delhi-110002 (INDIA)
Phones: +91-11-23243489, 43551243; Mobile: +91-9871775858
e-mail: katariabook@gmail.com; katariabooks@yahoo.com
Website: www.skkatariaandsons.com
(e-books, subscriptions and bulk buying also available on our website)

Published by:

S.K. Kataria & Sons®

4885/109, Prakash Mahal, Dr. Subhash Bhargav Lane,
Opposite Delhi Medical Association, Daryaganj, New Delhi-110002 (INDIA)
Phone: +91-11-23243489, 43551243; Mobile: +91-9871775858
e-mail: katariabook@gmail.com; katariabooks@yahoo.com
Website: www.skkatariaandsons.com

First Edition: 2010, Second Edition : 2019, Reprint: 2020, 2021, 2022

Price: ₹ 150/-

Laser Typesetting by :
Plus Computers, Delhi

Printed at :
New A.S. Printing Press, Delhi

Preface to the Second Edition

Engineers in mechanical engineering (Diploma holders) are responsible for controlling production and quality of the product on the shop floor as well as are responsible for production, planning and control. They are also required to supervise erection, installation and maintenance of equipment including material handling and undertake work study for better utilization of available resources in addition to lead a team of workers and motivate them towards realization of organizational objectives.

Our humble effort is to make them aware of the subject matter and enable them to develop competencies to face the situation. The book will teach them to understand and handle production environment effectively and efficiently. This edition in your hands has been tailored according to the revised syllabus of various polytechnics and covers almost all the topics for Operations Management being taught. It will also help in developing skills in erection, installation and testing of equipments and attachments. Study of the topics will also educate students of MBA immensely.

Despite our efforts to make the present version absolutely error-free, some errors and omission must have crept in unware. It is said that errors are invisible until they are printed. We shall feel obliged to our dear readers to get these brought to our notice for correction in the next print. Healthy criticism and constructive suggestions to improve the quality of the text shall be greatly appreciated.

Lastly the author owes a deep debt of gratitude to M/s S.K. Kataria & Sons for bringing out the book in such a wonderful print and get-up with speed and efficiency.

January, 2019

A.P. Verma
Author

Dedicated to My

Darling Grand-Daughters

Rini, Tanvi and Rhea

Who put Thoughts in My Brain and Ink in my Pen

CONTENTS

CHAPTER 1

INTRODUCTION TO OPERATIONS MANAGEMENT

INSIDE THIS CHAPTER

❒ Concept of Operations; ❒ Production and Operations Management ; ❒ Functions of Production and Operations Management; ❒ History of Development of Operations Management (OM); ❒ Scope and Application of OM; ❒ Benefits and Drawbacks of Operations Management; ❒ Types of Production (Manufacturing Systems); ❒ Necessity and Functions of Production, Planning and Control; ❒ Product Life Cycle.

1.1. CONCEPT OF OPERATIONS

For a long time, the term "Production" has been associated only with a factory like situation where goods are produced in the physical sense. A factory is any premises in which persons are employed for the purpose of making, altering, repairing, ornamenting, finishing, cleaning, washing, breaking, demolishing or adapting for sale, any article. However, by generalising the concept of production as the process through which goods and services are created, we can include both manufacturing and service organisations within the purview of production management. Thus, the essential features of the production function are to bring together people, machines and materials to provide goods and services to satisfy the wants of people. Inclusion of services within the scope of production enables us to look at the problem of production management in a much wider perspective. This brings a number of non-manufacturing sectors of economy such as transport, energy, health, agriculture, warehousing, banking and beauty-care etc., within the scope of production system. So, the term **Production and Operation Management or Operations Management** indicates the general applications of the techniques of management of machine and materials. The techniques used in the context of manufacturing organisations are always extended and applied to all types of service organisations as well.

A more general concept of "Operations" instead of "Production" will better include both manufacturing and service organisations. All operations add value to some object thereby enhancing its usefulness.

Definitions: An operation may be defined as **"The process of changing inputs into outputs and thereby adding value to some entity"**.

1.2. PRODUCTION AND OPERATIONS MANAGEMENT

Production and operations management concerns itself with the conversion of inputs into value-added outputs, (product/services) using physical resources, so as to

provide the desired utility ___ of form, place, possession or state or a combination thereof ___ to the customer while meeting the other organizational objectives of effectiveness, efficiency and adaptability.

Its functions are different from other functions such as personnel and marketing etc. by way of its primary concern for 'conversion by using physical resources'. But the use of other resources such as *'information'* in production and operation management can also not be denied. The management of the use of physical resources for the conversion process is what distinguishes production and operation management from other functional disciplines. The Table 1.1 illustrates the many facets of it.

Product and Services

The output of an operation system (or production system) may be in terms of end-product—physical goods such as automobiles, cement, paper, butter and books, etc. or rendering a service such as in transportation, hospitals, educational institutions, banks, cinema halls and beauty parlours, etc. Rendering a service may involve physical goods (or facilitating goods) such as dentist making a set of false teeth while rendering dental care, manufacturer of an automobile giving after-sale service to its purchaser after selling the automobiles and a beauty parlour providing a body massage cream made by them while doing a "facial service" to the customers. Thus, services can be considered as bundles of benefits, some may be tangible and others intangible (such as reduced waiting, courteous calls, convenient location, etc.) and these may or may not be accompanied by facilitating goods. Based on this grouping it is possible to segregate organizations producing goods or service or both.

Table 1.1. Production and Operations Management—Some Cases

Sl. No.	*Case*	*Input*	*Physical Resource/s Used*	*Output*	*Type of Input/Output*	*Type of Utility Provided to the Customers*
1.	Inorganic chemicals production	Ores	Chemical plant and equipment other chemicals, use of labour, etc.	Inorganic chemical	Physical input and physical output	Form
2.	Outpatient ward of a general hospital	Unhealthy patients	Doctors, nurses, other staff, equipment, other facilities	Healthier patients	Physical input and physical output	State
3.	Educational institution	Raw minds	Teachers, books, teaching aids, etc.	'Enlightened' minds output	Physical input and physical	State
4.	Sales office	Data from market	Personnel, office equipment and facilities, etc.	Processed 'information'	Non-physical input and non-physical output	State

(Contd...)

5.	Petrol pump	Petrol (in possession of the petrol pump owner)	Operators, err- and boys, equip- ment, etc.	Petrol (in possession of the car owner)	Physical input and physical output	Posse- ssion
6.	Taxi service	Customer (at railway station)	Driver, taxi itself, petrol	Customer (at his residence)	Physical input and physical output	Place
7.	Astrologer/ palmist	Customer (mind full of questions)	Astrologer, Panchanga, other books, etc.	Customer (mind with less questions, hopefully)	Physical input and physical output	State
8.	Maintenance workshop	Equipment gone 'bad'	Mechanics, Engi- neers, repairs equipment, etc.	'Good' Equipment	Physical input and physical output	State and form
9.	Income tax office	'Information'	Officers and other staff, office facility	Raid	Non-physical input and phy- sical output	State (posse- ssion ?)

1.2.1. System Concepts in Operations Management

Perhaps a more general concept of 'operations' instead of 'production' will better include both manufacturing as well as service organizations. Operations _____ either in manufacturing or in service _____ are purposeful activities of an organization. Operations function is the heart of and indeed the very reason for an organization to come into being as all operations add value to some object thereby enhancing its usefulness.

A system may be defined as "a purposeful collection of people, objects and procedures for operating within an environment". Thus, every organisation can be represented as a system consisting of interacting sub-systems. The features of a system are that these have inputs and outputs. The basic process of the system converts the resource inputs into some useful form of outputs. Of course, depending upon the efficiency of the conversion process we may have undesirable outputs too—such as pollution, scrap or wastage, rejections, loss of human life (in a hospital), etc. Using the generalised concept of production (which includes services) we can call such systems as production systems. A conceptual model is shown below.

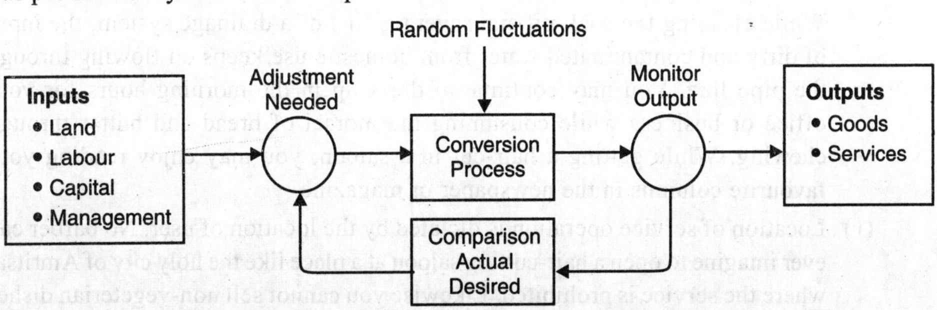

Fig. 1.1. Conceptual Model of a Production/Operation System (Elements of P/O System) as given by Adam and Ebert, 1978.

Figure 1.1 describes a generalised concept of production system. It takes resource inputs and processes them to produce useful outputs in the form of goods or services.

1.2.2. Types of Business Operation–Manufacturing, Service and Trading

As discussed earlier, the generalised model of production system includes both manufacturing system as well as service system. Examples of manufacturing system are : manufacturing of fertilizers, cement, paper, coal, textile, automobiles, machine tools, TV and furniture, etc. While that of service system include telephone exchange, post office, bank, hospital, beauty parlour, municipal corporation, transport organization, university, supply office and cinema halls, etc. Although basic structure of service system is amendable to the same analysis as manufacturing system, service systems do have the following salient features :

(*i*) Output from the service system is non-inventoriable. You cannot generally produce to stock the "aggregate of services". (No inventory is accumulated). It produces intangiable goods.

(*ii*) Quality of output is highly variable. For example, in a hair cutting saloon, customers may require the services of hair cutting according to their own choice and style (look at the hair-do of Michael Jackson, Gicketeer Chris Gayle and that of movie-director Rakesh Roshan) apart from multiple services of beard shaving and trimming, hair colouring and curling, etc. In a hospital, service is rendered right from the skin disease to the abdominal trouble to the kidney transplant and many more.

(*iii*) Service operations may be labour-intensive. The work of typing of a typist on a single foolscape page may cost you ₹ 6 or so (while the cost of one sheet is only 10 paise), the service of a facial in a beauty-parlour may cost you around ₹ 200 with only ₹ 5 worth of cream on your face, the ornamental work on a piece of cloth worth ₹ 10 may amount to as high as ₹ 1000/- approximately or even higher.

(*iv*) Production and consumption occur simultaneously. A manufactured good may be fully processed on all the machines and through the assembly lines, then inspected, okayed, distributed through the network of stockists and retailers and then finally reaches the hands of a customer or consumer. But the "stock" of the production of service and its consumption may occur simultaneously. While cleaning the soak-pit and severage line of a drainage system, the input of dirty and contaminated water from domestic use keeps on flowing through the pipe line. You may continue to dress up in the morning hours for your office or business while consuming the morsel of bread and butter through chewing. While getting a hair-cut in a saloon, you may enjoy reading your favourite columns in the newspaper or magazine.

(*v*) Location of service operation is dictated by the location of user. No barber can ever imagine to open a hair-cutting saloon at a place like the holy city of Amritsar where the service is prohibited. Likewise you cannot sell non-vegeterian dishes like fish, meat and chicken in and around the pilgrim place Rishikesh and umbrella and raincoats outlets can never be thought to be established in a desert area of Jaisalmer (Rajasthan).

(*vi*) Service (the intangible good) once rendered to the customer can never be taken back in case the customer does not like the quality of service whereas a tangible product made by a company, if not accepted in quality by the buyer can always be returned back.

That is why for the above reasons, the management of the service system is more difficult than that of the manufacturing system. No physical product, however, can be devoid of services. For example, when a customer buys a Kirloskar Genset, he does not purchase only a genset, but also the confidence of a reputed brand name that spells a reliable after-sale service and a guaranteed replacement of defective parts. With the accelerating growth of technological advances, chances are that the quality of the products of the competing manufacturers will be more or less the same, it will be the quality of service that will make the real difference.

Trading : It is an act or process of buying, selling or exchanging commodities at either wholesale or retail prices within the country or between countries.

1.2.3. Determinants of Service Quality

These are the determinants of service quality.

(*i*) Reliability (consistency of performance and dependability).

(*ii*) Responsiveness (timely and prompt response to the service need of the customer).

(*iii*) Competence (knowledge and skill of the contact personnel and the support personnel).

(*iv*) Access (convenience of location ease of approachability and contact and reasonable working hours).

(*v*) Courtesy (politeness and friendliness of attendants and consideration for the customer).

(*vi*) Communication (customer is properly informed about the service, agreed terms and staff should listen to and understand him).

(*vii*) Credibility (trust worthiness of the service company, its reputation and brand name).

(*viii*) Security (physical safety, financial security to the customer and confidentiality of the service rendered).

(*ix*) Tangibles (consists of physical aspects of service—the physical evidence of the service : physical facilities, equipment used and appearance of the people delivering the service).

1.2.4. Objectives of Production and Operations Management (Criteria of Performance for the Production and Operations Management)

Every system (or organisation) has a purpose, certain objectives and goals to achieve. Since the objectives of an organisation have hierarchical structure, **sub-goals lead to accomplishment of goals which contribute to the achievement of objectives and eventually the purpose or mission of an organisation.** It is very important that these objectives should be unambiguously identified, properly structured and explicitly stated.

In general terms, the objectives of an organisation may be to produce the goods or services in required quantities and of quality as per schedule and at a **minimum cost.** Thus, quantity, quality and time schedule are the objectives that determine the extent of customer satisfaction. If an organisation can provide for these at a minimum cost, then the 'value' of goods created or services rendered enhances and that is the only way to remain competitive. Thus, various objectives can be grouped as—performance objectives and cost objectives.

Performance Objectives

The performance objectives may include :

(*a*) **Efficiency** or productivity as output per unit of input.

(*b*) **Effectiveness :** It concerns whether a right set of outputs is being produced. Where efficiency may refer to 'doing things right', effectiveness may mean 'doing the right things'.

(*c*) **Quality :** Quality is the extent to which a product or service satisfies the customer needs. The output has to conform to quality specifications laid down before it can be accepted.

(*d*) **Lead times :** Manufacturing lead time or throughput time is the time elapsed in the conversion process. Minimisation of idle time, delays, waiting, etc. will reduce throughout time.

(*e*) **Capacity utilisation :** Percentage utilisation of manpower, machines, etc.

(*f*) **Flexibility :** If the conversion process has the flexibility of producing a combination of outputs, it is possible to satisfy a variety of customer needs.

Cost Objectives

Attaining **high degree of customer satisfaction** on performance front must be coupled with **lower cost of producing** the goods or rendering a service. Thus, cost minimisation is an important systems objective. Costs can be explicit (visible) or implicit (hidden or invisible). These could be tangible in economic terms or intangible in social cost terms—such as delayed supplies, customer complaints, etc. While managing production systems, we must consider both the visible and invisible, tangible and intangible costs. Some examples of these costs are:

(*a*) **Explicit (visible) costs :**

- Material cost
- Direct and indirect labour cost
- Scrap/rework cost
- Maintenance cost

(*b*) **Implicit (invisible/hidden) costs :**

- Cost of carrying intentory
- Cost of stockouts, shortages, back logging, lost sales
- Cost of delayed deliveries
- Cost of material handling
- Cost of inspection
- Cost of grievances, dissatisfaction
- Downtime costs
- Opportunity costs.

For the purpose of managerial decision-making, we should consider the total relevant systems costs including visible and invisible. A longer term cost implications rather than only short-term will help in arriving at better decisions.

1.3. FUNCTIONS OF PRODUCTION AND OPERATIONS MANAGEMENT

(*a*) **Planning :** Planning the conversion systems, planning the use of the conversion systems. Planning involves aims and objectives and action to achieve them in the organization.

(*b*) **Organising :** Identifying and grouping work to be done for conversion, structureing of operations, staffing, Job and work design, Production/Operations standards, Payment systems etc.

(*c*) **Directing and Controlling :** Monitoring and control of conversion system in respect of :

(*i*) Quantity
(*ii*) Quality
(*iii*) Time
(*iv*) Inventory
(*v*) Cost
(*vi*) Maintenance.

(Directing is concerned with carrying out the desired plans related with the target and controlling is exercising commands for achieving of objectives of the enterprise).

A schematic diagram of the above is shown in Fig. 1.2. It will be noted that some of the areas of decision are overlapping in nature.

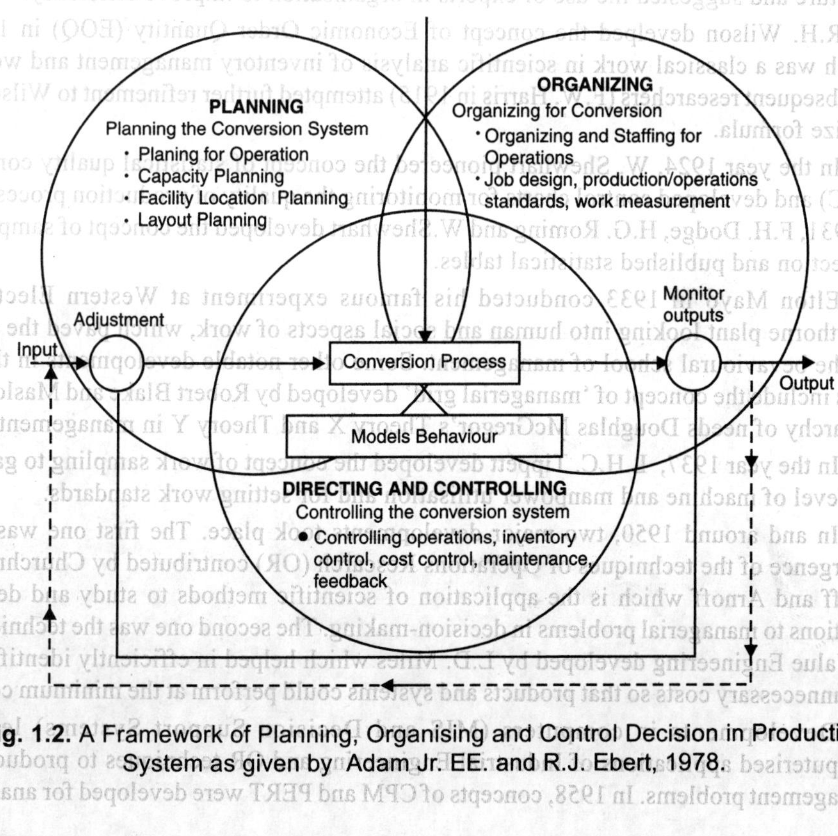

Fig. 1.2. A Framework of Planning, Organising and Control Decision in Production System as given by Adam Jr. EE. and R.J. Ebert, 1978.

1.4. HISTORY OF DEVELOPMENT OF OPERATIONS MANAGEMENT (OM)

The roots of OM go back to the concept of "division of labour" advocated by Adam Smith in his book " The Wealth of Nations" in 1776. In the year 1832, Charles Babbage, a mathematician extended Smith's work by recommending the use of scientific methods for analysing factory problems. The era of scientific management, however, started with the work of F.W. Taylor in 1878, who was subsequently recognised as "Father of Scientific Management", His experiments conducted on the shop floor brought about rapid increases in productivity. He is credited with recognising the potential improvement to be gained from analysing the work contents of a job and designing the job for maximum efficiency. Taylor described his management philosophy in his book, "The Principles of Scientific Management" in 1911 and and this event is considered as the beginning of the field of OM.

Frank Gilbeeth developed the concept of THERBLIGS and chronocyclegraphs for motion study in 1911 and his wife Lillian Gilbreth wrote her book "The Psychology of Management" which was one of the earliest works concerning the human factors in organisations.

In the year 1913, Henry Ford developed the concept of mass production and arranged work stations into an assembly line with moving belt. In the same year Henry Gantt made his visual diagrammatic tool which is popularly known as "Gantt Chart" and is an effective practical tool even today.

In 1913 also, Harrington Emerson applied Taylor's idea to develop organisation structure and suggested the use of experts in organisation to improve efficiency.

R.H. Wilson develped the concept of Economic Order Quantity (EOQ) in 1928 which was a classical work in scientific analysis of inventory management and works of subsequent researchers (F.W. Harris in 1915) attempted further refinement to Wilson's lot size formula.

In the year 1924, W. Shewhart pioneered the concept of statistical quality control (SQC) and developed control charts for monitoring the quality of production processes. In 1931, F.H. Dodge, H.G. Roming and W.Shewhart developed the concept of sampling inspection and published statistical tables.

Elton Mayo in 1933 conducted his famous experiment at Western Electrics Hawthorne plant looking into human and social aspects of work, which paved the way for the bevavioural school of management. Some other notable developments in these lines include the concept of 'managerial grid' developed by Robert Blake and Maslow's hierarchy of needs Doughlas McGregor's Theory X and Theory Y in management.

In the year 1937, L.H.C. Tippett developed the concept of work sampling to gauge the level of machine and manpower utilisation and for setting work standards.

In and around 1950, two major developments took place. The first one was the emergence of the techniques of Operations Research (OR) contributed by Churchman, Acoff and Arnoff which is the application of scientific methods to study and devise solutions to managerial problems in decision-making. The second one was the techniques of Value Engineering developed by L.D. Miles which helped in efficiently identifying the unnecessary costs so that products and systems could perform at the minimum costs.

Developments in computers (MIS and Decision Support Systems) led to computerised applications of Industrial Engineering and OR techniques to production management problems. In 1958, concepts of CPM and PERT were developed for analysis of large projects. In late 1960s the concept of OM extended to include the service sectors as well.

In the more recent past, there has been a major thrust to the adoption of Japanese Management techniques like Just in Time (JIT) system, Quality Circles (Q.C.) and Kanbon system for production scheduling and inventory control. Other notable developments in recent past have been Group Technology (GT), Cellular Manufacturing System (CMS), Flexible Manufacturing Systems (FMS) and Computer-aided Design/ Manufacturing (CAD/CAM), etc.

The future of OM looks bright in the years to come.

1.5. SCOPE AND APPLICATION OF OM

The following aspects are said to fall under the perview of OM:

1.5.1. Planning Physical Facilities

The factory building must be constructed to provide better working conditions to the employees. So, this involves provision of adequate physical facilities with the factory for lighting, ventilation, air-conditioning, sanitation and noise control. Planning must also be done for proper maintenance of plant machinery and other equipments. Finally industrial security is to be provided for men, materials, machinery and finished goods.

1.5.2. Organisation of Production

It implies all the factors of production such as raw materials, manpower, machines and tools, etc. required for production. Having procured these factors, necessary efforts are made to utilise them to the fullest extent possible to maximise production. The internal organisation of a factory helps in coordinating the efforts of all employees for the successful conduct of operations.

1.5.3. Factory Management

It is concerned with the implementation of policies into action. This requires the performannce of certain functions such as planning of the various aspects in the factory, namely—

(*i*) Directing various activities

(*ii*) Motivating the employees so as to increase their efficiency

(*iii*) Communication of various reports to all levels of management

(*iv*) Staffing the factory with required personnel

(*v*) Making decisions from alternative proposals

(*vi*) Coordinating the efforts of personnels.

1.5.4. Material Management

It is concerned with controlling the kind, amount, location, movement and timing of various commodities used by the factory. Effective control of materials embrace four phases:

(*i*) Procurement of materials

(*ii*) Receiving, traffic and shipping

(*iii*) Material handling

(*iv*) Storekeeping and inventory control

1.5.5. Labour Administration

It is concerned with the required number of employees with necessary skills for doing various types of work in the factory. Labour administration includes training of employees, promotions and transfer and performance rating of employees.

1.5.6. Production and Administration Control

The managerial function of control consists of emparison of actual performance with the planned performance to discover whether everything is proceeding as per plan and if not *why*. Effective control of men, materials and machines is essential for proper running of an enterprise.

1.6. BENEFITS AND DRAWBACKS OF OPERATIONS MANAGEMENT

1.6.1. Benefits of Operations Management

The efficient production and operations management gives us the following benefits:

(*a*) Consumer benefits from improved industrial productivity, increased used value of the product. Products are available at right place, right price and right time in desired quantity and quality.

(*b*) Investors get increased security for their investments, adaquate market returns and creditability and good image in the society.

(*c*) Employees get adaquate wages, job security, improved working conditions and increased personal and job satisfaction.

(*d*) Suppliers will get confidence in management and their bills can be realised without delay.

(*e*) Community enjoys benefits from economic and social stability.

Drawbacks

(*a*) If the quality control section of the enterprise does not exercise strict inspection and supervision work, then the cost of inspection and rework, scrap, spoilage and field complaints are much higher.

(*b*) The errors made during selection of materials, the selection of conditions under which the design has to function and the condition of supplied material may prove costly for the management as failure of the finished product can tarnish their image.

(*c*) Sometimes quality is compromised due to early delivery schedules, reduction in cost and safety, etc. Such activities in OM may bring down the name of the management in the market.

1.7. TYPES OF PRODUCTION (MANUFACTURING SYSTEMS)

Different types of production are shown in the following table:

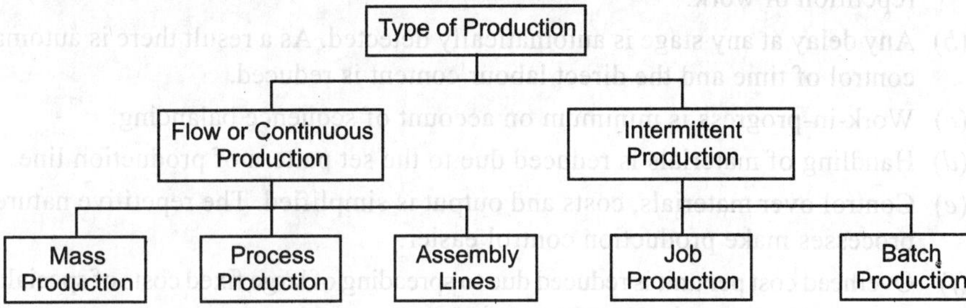

Fig. 1.3. Types of Manufacturing Systems

1.7.1. Continuous or Flow Production

Meaning. According to Buffa, **"continuous flow production situations are those where the facilities are standardised as to routing and flow since inputs are standardised. Therefore, a standard set of processes and sequence of processes can be adopted".** Thus, continuous or flow production refers to the manufacturing of large quantities of a single or at most a very few varieties of products with a standard set of processes and sequences. The processing is done through a sequence of operations which forms a continuous movement or flow. The work flows through a series of interconnected operations in such a way that the material passes from one stage to another without waiting or interruption. Under high volume, standardised production, the operations form a sequence and the conveyors connecting different operations themselves act as storage space. Steel mills, automobile plants, chemical plants, oil mills, sugar mills, etc., are examples of continuous production system.

Characteristics

The main features of continuous production system are as follows:

(*i*) The volume of output is generally large (mass production), and goods are produced in anticipation of demand.

(*ii*) The product design, the conversion process and the operations sequence are standardised, i.e., identical products are produced.

(*iii*) Machines and equipment are arranged according to product layout pattern.

(*iv*) Special purpose automatic machines are used to perform standardised operations.

(*v*) Machine capacities are balanced so that materials are fed at one end of the process and finished product is received at the other end.

(*vi*) Fixed path material handling equipment is used due to the predetermined sequence of operations.

Merits and Demerits

The main advantage of continuous system is that work-in-progress inventory is minimum. As the processing of material is continuous and progressive, there is no waiting period. Each work is passed to the next stage immediately after the previous operation is complete without waiting for the completion of work on the total batch. As a result costs of materials handling are minimised and full use can be made of automation. Few job instructions are needed and less storage space is required. Cost per unit can be minimised by investing in specialized equipments.

Continuous production offers the following advantages :

(*a*) The quality of output is kept uniform because each stage develops skill through repetition of work.

(*b*) Any delay at any stage is automatically detected. As a result there is automatic control of time and the direct labour content is reduced.

(*c*) Work-in-progress is minimum on account of sequence balancing.

(*d*) Handling of materials is reduced due to the set pattern of production line.

(*e*) Control over materials, costs and output is simplified. The repetitive nature of processes make production control easier.

(*f*) Overhead cost per unit is reduced due to spreading of large fixed costs of specialised equipments over a large volume of output.

(*g*) There is quick return on capital employed.

Continuous system, however, is very rigid and if there is fault in one operation, the entire process is disturbed. Due to continuous flow, it becomes necessary to avoid piling up of work or any blockage on the line. Unless the fault is cleared immediately it will force the preceding as well as the subsequent stages to be stopped.

Suitability

Continuous system is best suited to organisations which intend to produce a limited variety of products on a large scale. The heavy fixed costs of specialised equipments that are utilised for operating at low cost per unit can be distributed over a high volume of output. Continuous production system can be applied to those industries which satisfy the following requirements :

(*a*) **Uniform Demand.** The market demand for products should be stable and regular so that there is no piling up of stock. However, a provision should be made for balancing and adjusting demand. During peak demand, the production may be supplemented while during slack demand, the material may be stored. There should not be significant gap between the rate of output and the average level of demand.

(*b*) **High Volume of Production.** Unless the volume of output is large enough overhead costs of specialised machines may be very high.

(*c*) **Product Standardization.** The product must be standardised so as to permit the use of same process and material and a steady flow of output.

(*d*) **Process Balancing.** There must be a proper balance between different stages of operation of production so as to maintain a uniform flow of product. Raw materials should also be easily available.

(*e*) **Material Flow.** In order to maintain the flow of work, correct quantity and quality of material should be fed to the production process at the right time.

(*f*) **Preventive Maintenance.** The breakdown of one machine in the line results into total stoppage of the production process. Therefore, preventive maintenance should be undertaken to minimise downtime and without waiting for breakdown of machinery.

Continuous production is of the following types :

1. Mass Production

Mass production refers to the manufacture of standardised parts or components on a large scale. Standardisation of materials, machines, products and processes is the basic feature of mass production. The parts manufactured under mass production are combined in assembly line for making different products. Generally the degree of mechanisation and automation is high. The production process is continuous.

Mass production system is employed in many industries where the production is carried on without interruption. Electronics, electricals, automobiles, bicycles and container industries are a few examples of mass production industries. Mass production system offers economy of scale as the volume of output is large. Quality of products tends to be uniform and high due to standardisation and mechanisation. In a properly designed and equipped process, individual expertise plays a less prominent role. Of course the exact quality level depends upon the quality control system and management policy of the plant.

2. Process Production

Under this type of flow production, production is carried on continuously through a uniform and standardised sequence of operations. Highly sophisticated and automatic machines are used. Process production is employed in bulk processing of certain materials. For example, crude oil is processed into petroleum, kerosene, diesel oil and gasoline. Similarly, sugarcane is processed into sugar and molasses. In the automobile industry, process production is used in making chassis. Chemicals, cement and paper, etc. are other examples of process production industries. Plants employing process production system are known as '*process plants*'.

On the basis of the nature of production process, flow production may be classified into analytical and synthetic production. In **analytical process** of production, a raw material is broken into different products. For example, crude oil is analysed into gas, naphtha, petrol, etc. Similarly, coal is processed to obtain coke, coal gas, ammoniacal liquor, coaltar, etc. (Fig. 1.4)

Synthetic process of production, on the other hand, involves the mixing of two or more materials to manufacture a product. For instance, caustic soda, acids and perfumes, etc. are synthesised to manufacture soap, hydrogen, sulphur and oxygen are combined together to make sulphuric acid. (Fig. 1.5)

In *assembling process*—two or more components are combined to manufacture a finished product. Such process is employed in assembling automobiles, radio sets, televisions, bicycles, watches, cameras, etc.

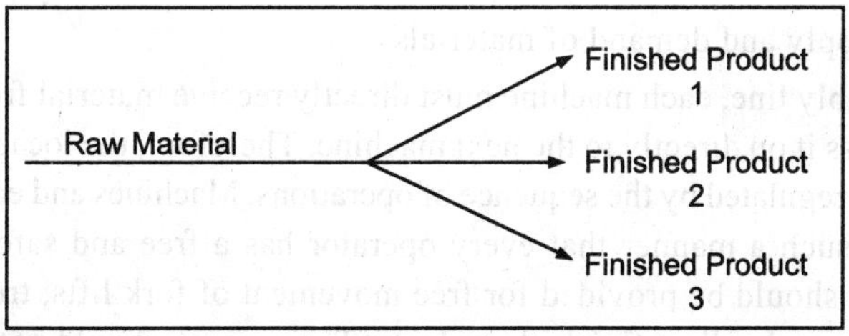

Fig. 1.4. Analytical Process.

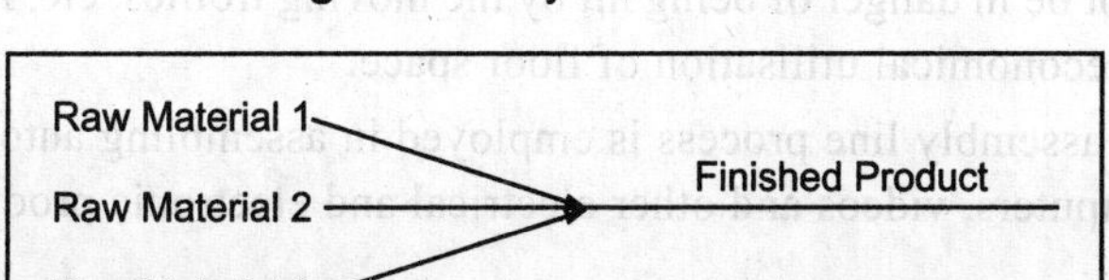

Fig. 1.5. Synthetic Process.

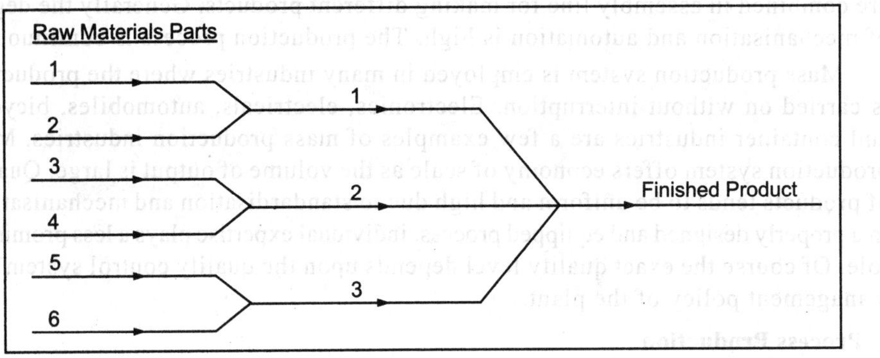

Fig. 1.6. Assembling process.

3. Assembly Lines

Assembly line is a type of flow production which developed in the automobile industry in the U.S.A. A manufacturing unit prefers to develop and employ assembly line because it helps to improve the efficiency of production. The use of flow production methods results in cost reduction. Assembly line is particularly useful when a limited variety of similar products is to be produced on a mass scale or in fairly large batches on a regular or continuous basis. (Fig. 1.6)

In any production system, the most vital decision is the proper layout of assembly line. The design of assembly line involves the proper balancing of technology and other manufacturing facilities so as to develop a rational approach for optimisation of results. The assembly line design depends largely upon product design and location of production. In order to develop an assembly line machines are positioned keeping in view the following considerations :

(*i*) The rate of flow of work.

(*ii*) The direction of manufacturing operations.

(*iii*) The convenience and comfort of operators or workers.

(*iv*) The availability of service facilities like water, electricity, compressed air, oxygen, etc.

(*v*) The supply and demand of materials.

In an assembly line, each machine must directly receive material from the previous machine and pass it on directly to the next machine. Therefore, the location of machines is automatically regulated by the sequence of operations. Machines and equipment should be arranged in such a manner that every operator has a free and safe access to each machine. Space should be provided for free movement of fork lifts, trucks, etc. which deliver materials and collect finished products. The passage should not be blocked and workers must not be in danger of being hit by the moving trollies, etc. At the same time there should be economical utilisation of floor space.

These days assembly line process is employed in assembling automobiles, radios, televisions, computers, videos and other electrical and electronic products.

1.7.2. Intermittent Production

According to Buffa, **"Intermittent production situations are those where the facilities must be flexible enough to handle a variety of products and sizes, or where**

the basic nature of the activity imposes change of important characteristics of the input (e.g., change in product design). In instances such as these, no single sequence pattern of operations is appropriate, so the relative location of the operation must be a compromise that is best for all inputs considered together."

Under intermittent production, the basic product design changes from time to time. Therefore, facilities should be flexible enough to handle a wide variety of products and of different shapes and sizes. Production facilities are arranged according to their functions so that functional skills and expertise can be concentrated on a specialised operation. Job type machine shops, batch type chemical operations are examples of intermittent model.

The main characteristics of intermittent production are as follows:

(*a*) The flow of production is intermittent, not continuous.

(*b*) The volume of production is generally small.

(*c*) A wide variety of products are produced.

(*d*) General purpose machines and equipment are used so as to be adaptable to a wide variety of operations.

(*e*) No single sequence of operations is used and periodical adjustments are made to suit different jobs or batches.

(*f*) Functional or process layout is adopted.

(*g*) The components are produced for inventory but they are assembled as per customer orders.

Intermittent system is much more complex than continuous production because every product has to be treated differently under the constraint of limited resources. Intermittent system can be effective in situations which satisfy the following conditions :

(*i*) The production centres should be located in such a manner so that they can handle a wide range of inputs.

(*ii*) Transportation facilities between production centres should be flexible enough to accommodate variety of routes for different inputs.

(*iii*) The skill and speed of each centre is different. In order to maintain a proper flow of work, it becomes, therefore, necessary to provide storage facility between different production centres. There must be sufficient flexibility of routes, volume and product design. Adequate storage between operations helps to make better utilization of men and machines.

The distinction between continuous and intermittent systems has nothing to do with the regularity of manufacturing operations. For example, a tomato cannery operates only a few months every year even though it is a case of continuous production. On the other hand, a job machine shop and custom made furniture factory, both intermittent, may operate throughout the year.

Intermittent Production may be of two types :

1. Job production ; and
2. Batch production.

1. Job Production (Produce to Order Manufacturing System)

Meaning. Job or unit production involves the manufacturing of a single complete unit with the use of a group of operators and processes as per the customer's order. This is a 'special order' type of production. Each job or product is different from the others and no repetition is involved. The whole job is one operation and work is completed separately on each job. The product is usually costly and non-standardised. There is absence of identical jobs and long runs on a single job. Jobs are carried out to the customers' specifications. Thus, job production consists of "the bringing together of materials, parts and components in order to assemble and commission a single piece of equipment or product." Under job production, goods are produced to specific customer orders. Customers do not make demand for exactly the same product on a continuing basis and, therefore, production becomes intermittent. Each product is a class by itself and constitutes a separate job for production purposes. Shipbuilding, electric power plant, furnace making, dams construction, jobbing functions, machine shops, printing press, etc. are common examples of job production. Firms engaged in subcontract, work such as tool making, machining, sheetmetal work and sub-assembly, use job production system.

Characteristics

The main characteristics of job production are as follows:

(*i*) The product manufactured is custom-made or non-standardised.

(*ii*) Machines and equipment are arranged or assembled at one place, *i.e.*, fixed position layout is used.

(*iii*) A wide range of general purpose machines like grinder, drill press, shaper, etc., is used.

(*iv*) Volume of output is generally small.

(*v*) Variable path materials handling equipments are used.

Merits

Job production involves comparatively small investment in machinery and equipment. It is flexible and can be adapted easily to changes in product design. A fault in one operation does not result into complete stoppage of the entire process.

Demerits

Job shop manufacturing is the most complex system of production. For instance, in building a ship, thousands of individual parts must be fabricated and assembled. A complex schedule of activities is required to ensure smooth flow of work without any bottlenecks. Many parts require operation time on the same machines even though the operations in the manufacturing cycle take place at different points of time. Optimum machine loading is, therefore, a very difficult problem.

Raw materials and work-in-progress inventories are high due to uneven and irregular flow of work. Therefore, large storage space is required and material handling costs are high. Highly skilled machine operators are required for the efficient performance of setting up and operating tasks. Workloads are unbalanced. Labour and equipment must be flexible to accommodate the continuously changing production.

There is often inefficient utilisation of labour and plant. Speed of work is slow and unit costs are high.

Thus, job shop production is applicable where custom-made products are to be produced on a small scale.

2. Batch Production

Meaning. Batch production has been defined as **"the manufacture of a product in small or large batches or lots at intervals by a series of operations, each operation being carried out on the whole batch before any subsequent operation is performed."** In this type of production, the work on any product is divided into few operations and each operation is completed for the whole lot before proceeding to the next operation. Each batch contains identical items but every batch is different from the other. A fixed quantity or batch rather than continuous supply is required. A particular operation on a batch is performed by one group and then it is passed on to other groups for subsequent operations. For example, the printing operation may be done by one group, the jobbing work by another, and so on. Generally, one group completes the operation on the whole lot before it is passed on to the next group for subsequent operation and a part of the lot is not transferred.

Thus, batch production is a mixture of mass production and job production. Under it machines turn out different products at intervals, each product being produced for comparatively short time using mass methods.

Industrial blowers, electric motors, book printing, export production, drugs, machine tools are usually done by batch system. In batch production, machines and other facilities become available for the production of another batch after the production of a particular batch is completed.

Both job production and batch production are similar in nature except that in batch production the quantity of product manufactured is comparatively larger.

Characteristics

The essential features of batch production are given below:

(*i*) A large variety of products are manufactured in lots or batches

(*ii*) Both general purpose machines (for producing parts) and special purpose machines (for assembling the parts) are used

(*iii*) Variable path material handling equipment is used

(*iv*) Machines and equipment are arranged according to the sequence of operations, *i.e.*, process layout is used.

Merits and Demerits

Batch production combines the features of both flow production and job production types. Some degree of specialisation is possible and capital investment is comparatively low. But work-in-progress inventory is high and large storage space is required. Due to frequent changes in product design no standard sequence of operations can be used. Machine set-ups and tooling arrangements have to be changed frequently.

The main problem in batch production is the idle time between one operation and the other. The work has to wait until a particular operation is carried out on the

Table 1.2. Difference between Production Control Systems

FACTOR	JOBBING	BATCH	MASS	PROCESS
	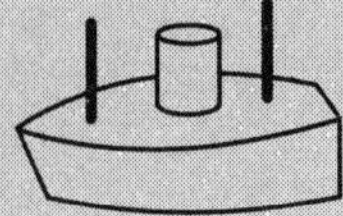	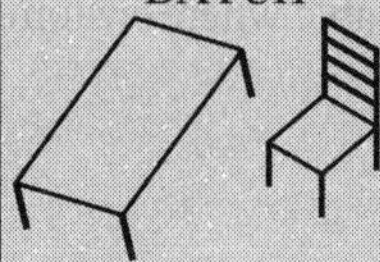	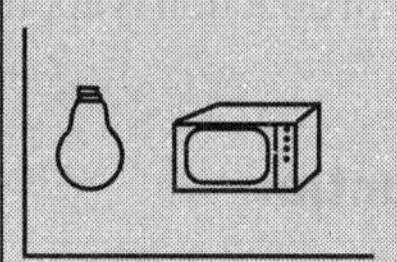	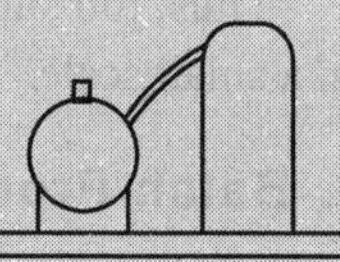
1. Equipment	Standard machinery size depending on whether the factory is engaged in light, medium or heavy engineering	Similar to Jobbing Production but there may be some special purpose equipment	Machinery designed for one range of product, largest product generally not greater than medium engineering	The entire 'factory' is completely integrated at all stages. Generally no isolated items of equipment
2. Type of building	In heavy and medium engineering will be single storey and either single or multistorey for light engineering. Floor area per worker will be high	Similar to Jobbing Production	Similarly to Jobbing batch Production to Batch Production the mass manufacture of Light engineering products will be single or multistorey buildings	Quite often the equipment in process manufacture will not be enclosed inside buildings
3. Layout of factory	Similar machines will be arranged in groups known as process or functional layout	Same as Jobbing Production but in some factories different machines may be grouped together to suit families of parts	All machines and processes will be arranged in operation sequence to suit the product-known as line Layout	The entire factory will be designed like one huge machine and to produce a certain rate of a specific product
4. Type of flow	Because of the difficulty of balancing demand with capacity, work will wait between operations-known as Intermittent Flow	Similar to Jobbing Production but a family grouping type of layout may reduce waiting but will also reduce machine utilisation	Demand and capacity will be reconciled so flow will be continuous	As the complete layout has been designed for a specified flow, production will be continuous
5. Cost and time required to make product	Cost of one product in relation to turnover will be high. Total time to make will be high and can be expressed as: $\frac{\text{Total operation time}}{\text{Total time to make}}$ This will always be less than 1	Similar to Jobbing Production but the ratio of total operation time to total time to make will be better	Cost of one product in relation to turnover will be small, The total time to make will be low and the ratio of total operation time to total time to make will be near to 1	Cost of one unit of output in relation to total output will be very small and the ratio of total operation time to total lime to make will be theoretically near to 1
6. Work in progress	The amount of W.I.P. in relation to total output will be high	Similar to Jobbing Production	The amount of W.I.P. in relation to total output will be small	There will theoretically be no W.I.P. between operation stages
7. Materials handling	Standard equipment and will depend on whether the factory is engaged in light, medium or heavy engineering	Similar to Jobbing Production but there may be some special purpose equipment	Extensive use of expensive specialised equipment	Materials handling and processing will be completely integrated
8. Foremen	Large technical content in Supervisory Function. Ratio of supervision to operators about 1:30	Specialists in the functions supervised e.g. Milling or Capstans Ratio of Supervision to operators about 1:50	Specialists in a particular aspect of flow production. Ratio of supervision to operators about 1:50	Concerned with co-ordinating and communicating. Ratio of supervision to operators about 1:15
9. Sales and design	Technical, large staff orders made to customers requirements	Product and Jig and Tool Design, Orders made to customer's requirements	Research and development, Product testing, Market Research to test customer's requirements	Research and development for products and processes, Market Research to test customer's requirements
10. Type of industry	Ship building, Civil Engineering, Process Equipment, Castings	Machine tools, Furniture, Clothes, Drop forgings	Food, cigarettes, T.V. electric lamps, vehicles	Chemicals, oil, petrol

whole batch. For example, assume that a particular batch is processed through four stages. The first unit of a batch undergoes the first operation within one week but it has to wait until four weeks when the first operation on the whole batch will be completed. Similarly, after each operation waiting or rest time is involved. Such delay results in substantial work-in-progress and considerable time lag in production. As a result investment is high and return on capital employed is low. Sometimes, a buffer stock may have to be maintained in order to ensure operating flexibility and effective use of resources. A proper balance must be arrived at between investment in material and investment in other resources.

Suitability

Batch production is applied when either the volume of output increases resulting in some repetitiveness or the market demand is not uniform throughout the year. In the latter case occasional discontinuity in production occurs as switching to other product becomes necessary.

An analysis of production systems shown pictorially in Table 1.2.

1.8. NECESSITY AND FUNCTIONS OF PRODUCTION, PLANNING AND CONTROL

See Chapter 7 *(Sec 7.1, Page 105 and Sec 7.3 Page 107).*

1.9. PRODUCT LIFE CYCLE

See Chapter 7 *(Sec 7.5, Page 126).*

MODEL QUESTIONS

1. What do you understand by "System Concept" ? Explain briefly.
2. Describe in short systems concept in operations management.
3. What are the characteristics of a system?
4. Why is it more difficult to increase productivity of a service system as compared to a production system?
5. Write down the objectives of production and operations management and explain each of these briefly.
6. What are the various determinants of service quality ? Describe each of these in short.
7. Write down the necessities of effective planning and control.
8. What are the various components of operations, planning and control ?
9. Explain continuous production and intermittent production being followed in industries.
10. Write short notes on the following :
    (*a*) Mass production.
    (*b*) Batch production
    (*c*) Job order production
    (*d*) Assembly lines
    (*e*) Process production
11. Write an essay on History of development of Operations Management.
12. Explain briefly the scope annd application of Operations Management.

□□□

CHAPTER 2

PLANT LOCATION, PLANT LAYOUT AND MATERIAL HANDLING

INSIDE THIS CHAPTER

- Plant Location Decision and its Importance
- Plant Layout—Introduction
- Material Handling

2.1. PLANT LOCATION DECISION AND ITS IMPORTANCE

Traditionally, location has always meant as industrial plant/factory location down the ages, however, the concept of plant location has now been generalised into that of facility location since the facility could include a production operation or service system. The term **plant** has been used as synonymous to a factory, manufacturing unit or assembly unit which could include fertiliser, steel, paper, cement, rice milling plants, textile, jute, sugar mills, rubber factories, breweries, refineries, thermal or hydro-electric nuclear power station, etc. However, with the enlarged scope of a facility, this term can now be used to refer to banks, hospitals, blood banks, fire stations, police stations, warehouses, godowns, depots, recreation centres and beauty parlours, etc.

Facility location decisions are strategic, long term and non-repetitive in nature. Without sound and careful location planning in the beginning itself, the new facility may pose continuous operating disadvantages, for the future operation. A right selection of a plant location can make and a wrong one can mar an organisation. Location decisions are affected by many factors such as the technology used, the capacity, the financial position, the work force required and also by economic, political and social conditions in the various localities. The efficiency, effectiveness, productivity and profitability of the facility are also affected by the location decision. The facilities location problem is concerned primarily with the best location depending on the appropriate criteria of effectiveness. The following are some of the important factors affecting plant location.

2.1.1. Factors Affecting Plant Location

1. Input Considerations
   (*a*) Material—Quantity, quality, cost and regular supply
   (*b*) Land—Site availability and costs, cost of construction, constructional regulations

   (*c*) Equipment—Availability and cost
   (*d*) Plant utilities—Gas, electricity, coal, water, etc. availability and cost
   (*e*) Labour—Availability, supply, skill, wage rates, unionization
   (*f*) Capital—Equity and debt potential, banking facilities

2. Processing considerations
   (*a*) Production analysis—Educational and research facilities
   (*b*) Process analysis—Engineering and consultancy
   (*c*) Forecasting and scheduling—Data resources and capabilities
   (*d*) Production control—Inventory storage and future expansion
   (*e*) Maintenance—Service and repair facilities
   (*f*) Cost control—Accounting and credit facilities
   (*g*) Presence of related industries.
3. Output considerations
   (*a*) Distribution—Distribution and storage facilities.
   (*b*) Transportation—Facilities and costs
   (*c*) Present and future market potential
   (*d*) Local rates and taxes
4. Other considerations
   (*a*) Community attitude towards industry and company
   (*b*) Public and community services—educational, recreational, housing, medical and cultural, etc.
   (*c*) Stockholder interests
   (*d*) Organizational decentralisation policies
   (*e*) Environmental standards—Air, water, zoning and building codes
   (*f*) Political situation

2.1.2. When Does a Location Decision Arise ?

A location decision arises due to the various reasons listed below :

1. It may arise when a new facility is to be established.
2. In some cases, the facility or plant operations and subsequent expansion are restricted by a poor site, thereby necessitating the setting up of the facility at a new site.
3. The growing volume of business makes it advisable to establish additional facilities in new territories.
4. Decentralisation and dispersal of industries reflected in the Industrial Policy resolution so as to achieve an overall development of a developing country, would necessitate a location decision at a macro level.
5. It could happen that the original advantages of the plant have been overweighed due to new developments.
6. New economic, social, legal or political factors could suggest a change of location of the existing plant.

Whenever the plant location decision arises, it deserves careful attention because of the long-term consequences and any mistake in selection of a proper location could prove to be costly. Poor location could be a constant source of:

(*i*) higher cost

(*ii*) higher investment

(*iii*) difficult marketing and transportation

(*iv*) dissatisfied and frustrated employees and consumers

(*v*) frequent interruptions of production

(*vi*) abnormal wastages

(*vii*) delays in committed dates of supply

(*viii*) substandard quality

(*ix*) denied advantages of geographical specialisation and so on.

Once a facility is set up at a location, it is very difficult to shift later to a better location because of numerous economic, political and sociological reasons.

Economic reasons could include total costs, profits, availability of raw materials, labour, power, transportation facilities and market, etc.

Social reasons could include employee welfare and employment opportunities, etc.

Political reasons could be because of pursuance of a policy of decentralisation regional and developmental planning especially in a developing country like India. There could be security considerations on the risk of military invasions, sabotage from anti-social elements, etc. and some may be prone to natural calamities like floods, droughts, and earthquake, etc. Policy matters like anti-pollution, etc. would have to be given their due consideration.

Weber's Analysis

Alfred Weber's analysis was one of the first attempt to base location decisions on the transport cost of the raw material which was least mobile. He categorised raw materials into : **(*a*) Ubiquities**—to denote those raw materials which are available almost everywhere like sand, soil and water, etc., and **(*b*) Localised materials** having specific locations. These are further divided into pure material which contributes nearly the total weight of it to the finished goods and gross material which contributes only a small fraction of total weight to the finished goods. It is obvious that ubiquities hardly influence the decision of location. Weber formulated the following material index :

$$\text{Material Index (M.I.)} = \frac{\text{Weight of the localised material used in the finished product}}{\text{Weight of the finished product}}$$

If M.I. > 1, then location should be nearer to the source of the raw material

If M.I. ≤ 1, then the location should be nearer to the market.

The commonsense involved in the above conclusion is unquestionable but such an approach tacitly assumes the existence of a static point of the lowest transportation cost for the raw material.

2.1.3. Steps in the Facility Location Study

In the following two phases, the location studies are usually made:

(*i*) the general territory selection phase

(*ii*) the exact site/community selection phase among those available in the general locate.

The considerations vary at the two levels, though there is substantial overlap as shown in the Table 2.1.

Table 2.1. Overlap of Considecltiy Location.

	Location factors	*Phase I General territory selection*	*Phase II Particular selection of site and community*
1.	Market	*	—
2.	Raw materials	*	—
3.	Power	*	*
4.	Transportation	*	*
5.	Climate and fuel	*	—
6.	Labour and wages	*	*
7.	Laws and taxation	*	*
8.	Community services and attitude	—	*
9.	Water and waste	—	*
10.	Ecology and pollution	—	*
11.	Capital availability	*	*
12.	Vulnerability to enemy attack	*	*

An interdisciplinary team consisting of economists, accountants, geographers, town planners, lawyers, marketing experts, politicians, executives, industrial engineers, defence analysts and ecologists, etc. should be set-up for undertaking location studies.

Phase I. Territory Selection

The following are some of the important factors that influence the selection decision for the general territory/region/area selection:

(*a*) **Markets.** There has to be some customer/market for your product/service. The market growth potential and the location of competitors are important factors that could influence the location. Locating a plant or facility near to the market is preferred if promptness in service is required, if the product is fragile, or is susceptible to spoilage like glass articles, bangles, ceramic goods, fresh breads, jam, jelly, pickles, etc. Moreover, if the product is relatively inexpensive and transportation costs add substantially to the cost, a location close to the market is desirable. Assembly type industries also tend to locate near markets.

(*b*) **Raw Materials and Supplies.** Sometimes accessibility to vendors/suppliers of raw materials, part supplies, tools, equipment, etc. may be very important. The issue here is promptness and regularity of delivery and inward freight cost minimization.

If the raw material is bulky or low in cost, or if it is greatly reduced in bulk viz. transformed into various products and by-products, if it is perishable and processing makes it less so, then location near raw materials sources is important. If raw materials come from a variety of locations, the plant/facility may be situated so as to minimise total transportation costs. The costs vary depending upon specific routes, mode of transportation and specific product classifications.

Transportation Facilities. Adequate transportation facilities are essential for the economic operation of a production system. For companies that produce or buy heavy bulky and low value per ton commodities, water transportation could be an important factors in locating plants. It can be seen that civilisations grew along rivers/waterways etc. Many facilities/plants are located along river banks.

Manpower Supply. The availability of skilled manpower, the prevailing wage pattern, living costs and the industrial relations situation influence the location.

Infrastructure. This factor refers to the availability and reliability of power, water, fuel and communication facilities in addition to transportation facilities.

Legislation and Taxation. Factors such as financial and other incentives for new industries in backward areas or no-industry-district centres, exemption from certain state and local taxes, octroi, etc., are important.

Climate. Climatic factors could dictate the location of certain type of industries like textile industry which requires high humidity zones.

Phase II. Site/Community Selection

Having selected the general territory/region, next we would have to go in for site/ community selection. Let us discuss some factors relevant for this stage:

Community Facilities. These involve factors such as quality of life which in turn depends on availability of facilities like schools, places of worship, medical services, police and fire stations, cultural, social and recreation opportunities, housing, good streets and good communication and transportation facilities.

Community Attitudes. These can be difficult to evaluate. Most communities usually welcome setting up of a new industry especially since it would provide opportunities to the local people directly or indirectly. However, in case of polluting, or 'dirty' industries, they would try their utmost to locate them as far away as possible. Sometimes because of prevailing law and order situation, companies have been forced to relocate their units. The attitude of people as well as the state government has an impact on industrial location.

Waste Disposal. The facilities required for the disposal of process waste including solid, liquid and gaseous effluents need to be considered. The plant should be positioned so that prevailing winds carry any fumes away from populated areas and so that waste may be disposed off properly and at reasonable expense.

Ecology and Pollution. These days there is a great deal of awareness towards maintenance of natural ecological balance. There are quite a few agencies propagating the concepts to make the society at large more conscious of the dangers of certain avoidable actions.

Site Size. The plot of land must be large enough to hold the proposed plant and parking and access facilities and provide room for future expansion. These days a lot of industrial areas/parks are being earmarked in which certain standard sheds are being provided to entrepreneurs (especially small scale ones).

Topography. The topography, soil structure and drainage must be suitable. If considerable land improvement is required, low priced land might turn out to be expensive.

Transportation Facilities. The site should be accessible by road and rail preferably. The dependability and character of the available transport carriers, frequency of service and freight and terminal facilities are also worth considering.

Supporting Industries and Services. The availability of supporting services such as tool rooms, plant services, etc. need to be considered.

Land Costs. These are generally of lesser importance as they are non-recurring and possibly make up a relatively small proportion of the total cost of locating a new plant. Generally sperking, the site will be in a city, suburb or country location. In general, the location for large-scale industries should be in rural areas, which helps in regional development also. It is seen that once a large industry is set up (or even if a decision to this effect has been taken), a lot of infrastructure develops around it as a result of the location decision. As for the location of medium scale industries, these could be preferably in the suburban/semi-urban areas where the advantages of urban and rural areas are available. For the Small Scale Industries, the location could be urban areas where the infrastructural facilities are already available. However, in real life, the situation is somewhat paradoxical as people, with money and means, are usually in the cities and would like to locate the units in the city itself.

Some of the industrial needs and characteristics that tend to favour each of these locales are now discussed. **Requirements** govern choice of a **city location** are :

1. Availability of adequate supply of labour force.
2. High proportion of skilled employees.
3. Rapid public transportation and contact with suppliers and customers.
4. Small plant site or multi-floor operations.
5. Processes heavily dependent on city facilities and utilities.
6. Good communication facilities like telephone, telex, post offices.
7. Good banking and health care delivery systems.

Requirements governing the choice of a **suburban location** are :

1. Large plant site close to transportation or population centre.
2. Free from some common city building zoning (industrial areas) and other restrictions.
3. Freedom from higher parking and other city taxes, etc.
4. Labour force required resides close to plant.
5. Community close to, but not in large population centre.
6. Plant expansion easier than in the city.

Requirements governing the choice of a **country/rural location** are :

1. Large plant site required for either present demands or expansion.
2. Dangerous production processes.
3. Lesser effort required for anti-pollution measures.
4. Large volume of relatively clean water.

5. Lower property taxes, away from Urban Land Ceiling Act restrictions.
6. Protection against possible sabotage or for a secret process.
7. Balanced growth and development of a developing or underdeveloped area.
8. Unskilled labour force required.
9. Low wages required to meet competition.

2.2. PLANT LAYOUT—INTRODUCTION

Layout refers to the arrangement of facilities in a particular workstation. It may be plant layout, office layout, auditorium layout or retail store layout.

Plant layout implies the physical arrangement of machines, equipment and other industrial facilities on the factory floor in such a manner that they may be handled efficiently. It may be defined as **"Plant layout is the optimum arrangement of different facilities including man, machine, equipment, and material, etc., showing the space allocated for material movement, storage and all supporting activites from the receipt of raw material to the shipping of the finished goods for an overall economy of production."** It involves a judicious arrangement of production facilities so that work flows in, as direct a path as possible. According to the great management expert Riggs, the overall objective of plant layout is to design a physical arrangement that most economically meets the required output, quantity and quality. An optimum layout would permit an uninterrupted flow of work through the work station. It ensures best possible utilization of machines, workers and space. A plant layout study is required to create an arrangement that will minimise unit production costs. Such a study involves a careful analysis of all factors affecting layout. It is an important decision as it represents a long term commitment. It is also important because it affects the flow of materials and processes, labour efficiency, supervision and control use of equipment, use of space, expansion possibilities, etc. Plant layout covers not only the initial layout of machines and other facilities but encompasses revision or improvement in existing layout in the light of changes in the methods and techniques of production.

2.2.1. Objectives and Advantages

Some of the important objectives of a good plant layout are as follows :

1. Overall simplification of production process in terms of equipment utilization, minimisation of delays, reducing manufacturing time and better provisions for maintenance.
2. Overall integration of men, materials, machinery, supporting activities and any other considerations in a way that result in the best compromise.
3. Minimisation of material handling cost by suitably placing the facilities in the best flow sequence.
4. Saving in floor space, effective space utilization, and less congestion/confusion.
5. Increased output and reduced inventories-in-process.
6. Better supervision and control.
7. Worker convenience, improved morale and worker satisfaction.
8. Better working environment, safety of employees and reduced hazards.
9. Minimisation of waste and higher productivity.
10. Avoid unnecessary capital investment.
11. Higher flexibility and adaptability to changing conditions.

2.2.2. Evidence of Poor Layout

The following are the evidences of a poor layout:

1. Slow movement of materials through the plant.
2. High handling and transportation costs.
3. Crowded stock rooms and production departments.
4. Congested aisles and workplaces making them safety hazardous.
5. Inconveniently located service departments.
6. In-process materials are frequently damaged, misplaced or lost.

2.2.3. Principles of Plant Layout

While designing the layout of a plant, the following principles (essentials) should be followed:

1. **Principle of Overall Integration :** All the plant facilities are integrated one single operating unit considering all the servicing necessary for operations.
2. **Principle of Minimum Distance Moved :** Attempts are made to minimise movements as far as possible. However, the principle of overall integration should be considered while planning for the minimum distance moved. In fact weight × distance moved should be the minimum.
3. **Principle of Flow :** A layout must arrange the work areas as far as possible, in the same way as the sequence of operations or processes. The basic idea is to move the work towards completion progressively without interference of backtracking with a minimum of interruption or congestion.
4. **Principle of Cubic Space :** Economy is obtained by using effectively available space—both vertical and horizontal.
5. **Principle of Satisfaction and Safety :** Satisfaction of working people is an important factor to be kept in mind. Similarly due consideration must be given to safely requirements.
6. **Principle of Flexibility :** The layout must be sufficiently flexible to take care of the necessity of re-arranging production facilities, when the need comes.

2.2.4. Types of Plant Layout

There are four broad categories of plant layout :

1. Product layout (Sequence grouping of machines)
2. Process layout (functional layout or type grouping of machines)
3. Fixed position layout (Project layout)
4. Group layout (Cellular layout or Combination layout)

2.2.4.1. Product Layout

In product layout, machines and equipment are arranged in the sequence of the manufacturing operations required for the product. The materials is moved from one work station to another sequentially without any backtracking or deviation. It is called line layout because machines are mostly arranged in a sraight line. The raw materials are fed at one end and taken out as the finished product on the other. It is a grouping of machines in one sequence. A product layout may however, assume a straight line shape, a U-shape or circular shape. A straight line type of product layout is shown below in Fig. 2.1.

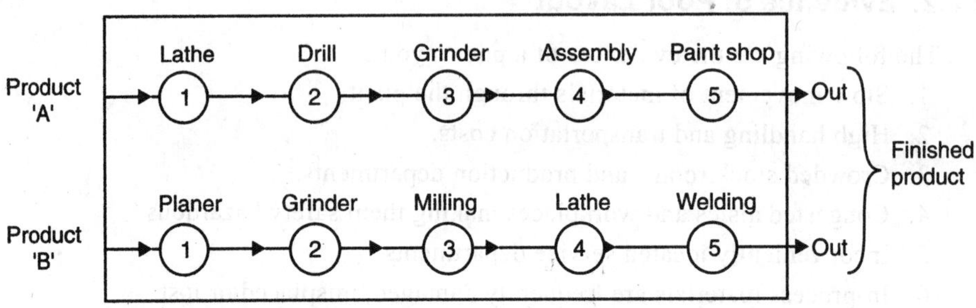

Fig. 2.1. Product Layout Showing Movement of Two Products.

Advantages. Product layout offers the following benefits :

(*i*) Low cost of material handling due to straight and short path and elimination of back tracking.

(*ii*) Smooth and uninterrupted operations from free bottlenecks.

(*iii*) Continuous flow of work permits mechanised handling of materials.

(*iv*) Lesser investment in inventory and work-in-progress.

(*v*) Special purpose equipment can be operated by semi-skilled labour.

(*vi*) Optimum use of floor space and less congestion of work in the process.

(*vii*) Shorter processing time or quicker output.

(*viii*) Simple and effective inspection of work and simplified production control.

(*ix*) Lower cost of manufacturing per unit.

Disadvantages. Product layout suffers from the following drawbacks :

(*i*) High initial capital investment in special purpose machines.

(*ii*) Heavy overhead charges.

(*iii*) Breakdown of one machine results in serious work stoppage of the entire line of machines.

(*iv*) Less flexibility of operations as layout cannot be adapted to the manufacture of any other type of product. Fluctuations in rate of production will increase manufacturing cost. Even a minor change in machine arrangement requires a complete change in layout.

Suitability. Product layout is useful under the following conditions :

(*i*) Mass production of standardised products.

(*ii*) Simple and repetitive manufacturing processes.

(*iii*) Operation time for different processes is more or less equal.

(*iv*) Reasonably stable demand for product.

(*v*) Interchangeability of parts.

(*vi*) Continuous supply of materials.

Large automobile assembly plants, food-processing chains and continuous process industries are examples where line layout is useful. For instance, "a multiple spindle automatic screw machine successfully indexes the work from one matching position to another so that several operations are performed in a continuous flow of material within the machine.

To conclude, line layout best suits those industries which manufacture a large volume of standard products involving repetitive processes. In these industries, line layout reduces inventory handling and supervision costs.

2.2.4.2. Process or Functional Layout

In this type of layout shown in Fig. 2.2 machines of similar type are located together according to their function. Products move between the groups of equipments in order of the operations required. For example, all the lathe machines are placed in the lathe department, all the drilling machines in the drilling department, all the welding equipment in welding department and so on.

Advantages. Process layout has the following merits :

(*a*) Lower initial capital investment is required in machines and equipment. There is higher degree of machine utilisation as a machine is not tied to a single product.

(*b*) Overhead costs are relatively low.

(*c*) Change in product design and volume can be more easily adapted to the output of a variety of products.

(*d*) Breakdown of one machine does not result in complete work stoppage.

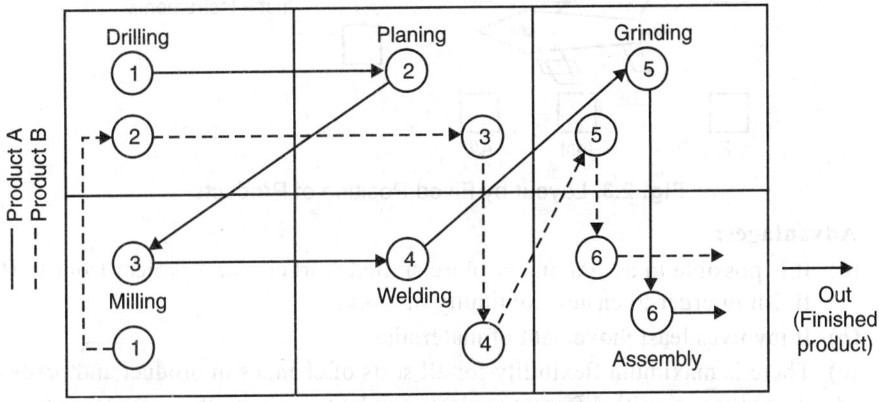

Fig. 2.2. Process Layout Showing Movement of Two Products

(*e*) Supervision can be more effective and specialised.

(*f*) There is greater flexibility and scope for expansion.

(*g*) Setup and maintenance costs are low.

Disadvantages. Process layout suffers from the following limitations :

(*a*) Material handling costs are high due to backtracking.

(*b*) More skilled labour is required resulting in higher labour cost.

(*c*) Processing time or time-lag in production is higher.

(*d*) Work-in-progress inventory is high requiring greater storage space.

(*e*) More frequent inspection is needed resulting in costly supervision. As the work has to pass through different departments, it is quite difficult to definitely trace the final responsibility for the finished product.

(*f*) Production planning and control become difficult.

Suitability. Process layout is useful when :

(*a*) Products are not standardised.

(*b*) Quantity produced is small.

(*c*) There are frequent changes in design and style of product.
(*d*) Job shop type of work is done.
(*e*) Machines are very expensive.

2.2.4.3. Fixed Position Layout

- Layout by fixed position of the product is inherent in ship building, aircraft manufacture (Fig. 2.3) and big pressure vessels fabrication.
- In other types of layouts discussed earlier, the product moves past stationary production equipment, whereas in this case the reverse applies; **In this fixed position layout, men and equipment are moved to the material, which remains at one place and the product is completed at that place where the material lies.**

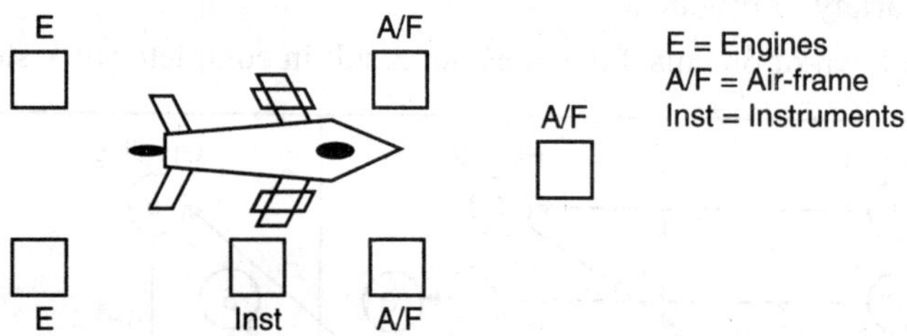

Fig. 2.3. Layout by Fixed Position of Products

Advantages:

(*a*) It is possible to assign one or more skilled workers for a project from start to finish in order to ensure continuity of work.
(*b*) It involves least movement of materials.
(*c*) There is maximum flexibility for all sorts of changes in product and process.
(*d*) A number of quite different projects can be taken with the same layout.

Disadvantages:

(*a*) It usually involves a low content of work-in-progress.
(*b*) There appears to be low utilization of labour and equipment.
(*c*) It involves high equipment handling costs.

2.2.4.4. Group Layout (Cellular Layout or Combination Layout)

A combination of process and product layouts combines the advantages of the both types of layouts. Moreover, these days pure product or process layouts are rare. Most of the manufacturing sections are arranged in process layout with manufacturing lines occurring here and there (scattered) wherever the conditions permit. **A combination layout is possible where an item is being made in different types and sizes. In such cases, machinery is arranged in a process layout but the process grouping (a group of number of similar machines) is then arranged in a sequence to manufacture various types and sizes of products.** The point to note is that, no matter the product varies in size and type, the sequence of operations remains same or similar. Figure 2.4 shows a combination type of layout for manufacturing different sized gears.

A combination layout is also useful when a number of items are produced in same sequence but none of the items are to be produced in bulk and thus, no item justifies for an individual and independent production line. For example, files, hacksaws, circular metal saws, wood saws, etc. can be manufactured on a combination type of layout.

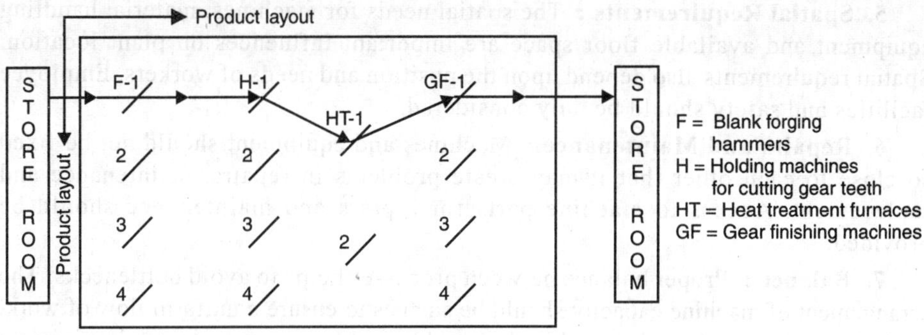

Fig. 2.4. A Combination Layout for Making Different Types and Sizes of Gears

2.2.5. Factors Affecting Plant Layout

The following factors should be considered while planning the layout of a plant:

1. Plant Location : Plant layout is intimately connected with plant location. As a matter of fact the size and shape of the site and its topography influence the general pattern of layout. Plant site also influences the type of building, mode of transport and the scope of expansion which in turn influences layout. Plant layout is generally a compromise between the ideal layout and the limitations of plant site and building. A decision to relocate offers an opportunity to improve the arrangement of facilities and services. Thus, plant layout is considerably affected by both the specific site and the general location.

2. Nature of Product : Standardised products require a product layout, whereas custom-made products may need process layout. A fixed position layout would be required for heavy and bulky products. Products involving hazardous and dangerous operations would require isolation of processes. Similarly, value, fragility, volume and quality of product are important considerations in plant layout.

3. Type of Industry : The nature and type of production process exert considerable influence on plant layout. Generally, product layout is more appropriate in continuous process industry, whereas intermittent production requires process layout. The sequence of operations should also be taken into consideration. Perhaps no one factor influences plant layout as much as the nature of the productive process involved. A synthetic process requires a line layout so that various materials and parts merge into the completed assembly in a flow. On the other hand, an analytic process takes the form of a tree as it starts out with a single material and spreads out into a variety of resultant materials. A conditioning process requires a layout to accommodate lot or batch manufacture. In such *a* layout, the material undergoes no merging or separation but it is simply subjected to physical conditioning, as it flows in separate lots through the variety of operations and processes.

Thus, the layout designer should keep in mind the type of industry and the nature of the manufacturing process while planning a layout.

4. Plant Environment : In planning factory layout, heat, light, noise, ventilation and other aspects of plant climate should be given due consideration. For example, paint shops and plating sections should be located on an outside wall so that dangerous fumes may be removed through proper ventilation. Type of machines, materials and equipment used also exercise considerable influence on plant location.

5. Spatial Requirements : The spatial needs for machines, material handling equipment and available floor space are important influences on plant location. Spatial requirements also depend upon the position and needs of workers. Employee facilities and safety should be duly considered.

6. Repairs and Maintenance : Machines and equipment should not be fixed so close to each other that it may create problems in repairs, maintenance and replacement. Access to machine parts for repairs and maintenance should be provided.

7. Balance : Proper balance between processes helps to avoid bottlenecks. The arrangement of machine capacity should be such as to ensure a uniform flow of work. At the same time the layout should be designed in such a manner that there is minimum possible movement of materials and men.

8. Management Policy : Management policies regarding size, quality, employee facilities and delivery schedules should be considered while deciding plant layout.

2.2.6. Plant Layout Procedure

The ideal procedure for a plant layout is to build the layout around the productive process and then design the building around the layout. This may not be possible always because the plant building may already be existing or the shape of the plant site may not permit the construction of a building to house the productive process, etc. Ultimately one has to strike a blanace between the two approaches.

A Systematic Layout Planning (SLP) procedure suggested by Francis and White (1974) has been shown in the Fig. 2.5.

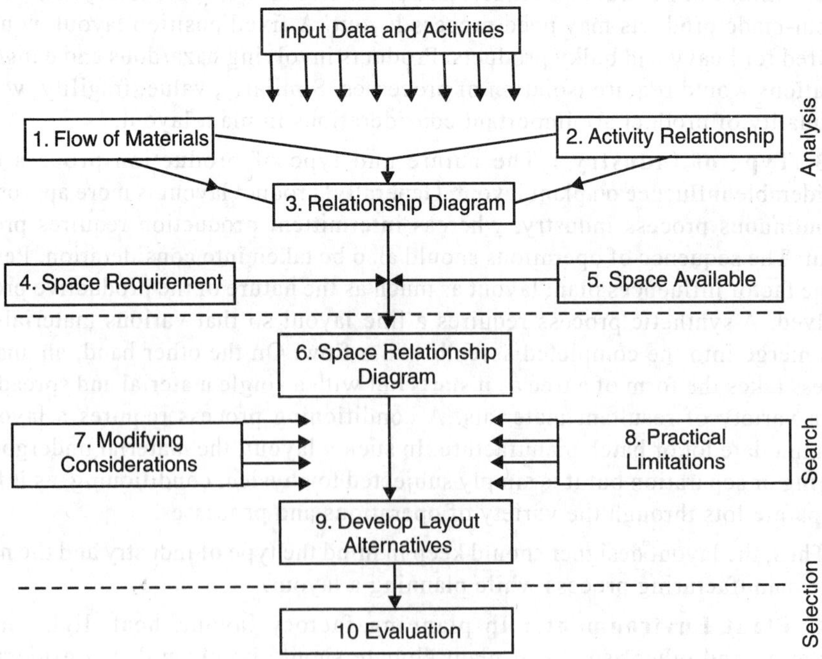

Fig. 2.5. Systematic Layout Planning Procedure

2.2.7. Line Balancing

"Line Balancing in a layout means arrangement of machine capacity to secure relatively uniform flow at capacity operation."

It can also be said as a "layout which has equal operation times at the successive operation in the process as a whole".

Product layout requires line balancing and if any production line remains unbalanced machinery utilization may be poor. The machine in the line will operate only say half of the time.

Let us assume that there is a production line with work stations, *x*, *y* and *z*. Also assume that each machine at *x*, *y* and *z* can produce 200 items, 100 items and 50 items per hour respectively. If each machine was to produce only 50 items per hour, then each hour the machines at *x* and *y* would be idle for 45 and 30 minutes respectively while machine *z* will be fully loaded. Such a layout will be unbalanced one and the production lines need balancing.

As an another example, a bakery would not be in balance, if the oven continuously baked loaves at the rate of 600 per hour, slicing takes place at the rate of 500 an hour and wrapping machine could only wrap 400 loaves per hour. Hence, the production line requires balancing for these mismatched activities.

2.2.7.1. Methods of Line Balancing

(*i*) The one possibility in the right direction, at least as far as balancing the line is concerned, would be to increase the output.

(*ii*) The second possibility is that another product may be made close to the first one so that some idle machines may be used jointly.

(*iii*) The third possibility may be to estimate the output of the last work station. This can be taken as the minimum output of all the intermediate work stations.

2.2.7.2. Objectives of Line Balancing

A balanced layout eliminates bottle-neck operations as well as prevents the unnecessary duplication of equipment capacity. Line balancing is a major consideration in layout because a lack of balance can most easily hinder the production. For balancing it is not essential that output of each operation should be the same but the essential is to see that output of fastest machines should be multiple of the output of the remaining other machines. The following example aims to illustrate line balancing:

Illustration : A product requires 4 operations on 4 work stations in the line. The operation time per unit is as follows :

Operation	Time per unit in minutes
1	10
2	20
3	10
4	20

The total working hours per month are 150. The required production is of 450 units per month. Suggest how should the time be balanced for maximum utilisation of machines. What percentage of each machine time is idle ?

Solution. Working hours/month = 150 hours = 150 × 60 = 9000 minutes

Production/month = 450 units

$$\text{Time/Unit} = \frac{9000}{450} = 20 \text{ minutes}$$

In the question operation time for the last operation 4 is 20 minutes and on the basis of this last operation work station utilisation can be calculated.

In this question work station 2, *i.e.*, operation 2 also consumes 20 minutes. From the above calculations in a production of 450 units per month, time per unit will be 20 minutes. Therefore, work stations 2 and 4 are 100% engaged while work stations 1 and 3 are 50% engaged. Work stations 1 and 3 can be loaded with other jobs taking approximately 10 minutes time each.

Heuristic (Here-you-stick) Method of Line Balancing. The heuristic method involves drawing a precedence diagram in a particular way which indicates the *flexibility* available for transferring tasks laterally from one column to another to arrive at the most promising balance. Heuristic produced very good results when Wester and Kilbridge applied it to T.V. assembly line problems (with 133 steps).

The heuristic approach involves following procedural steps : **(Wester and Kilbridge Method)**

1. Identify the work (job).
2. Breakdown the work into elemental tasks or steps. For example, drilling a hole is a step of elemental task.
3. List the various steps as under (An illustration) : It is required to design an assembly line for a cycle time of 10 minutes for the following 9 work elements whose immediate predecessors and duration times of each element are given below :

Table 2.2.

Steps or Elemental Tasks	*Immediate Predecessor*	*Duration of the Task (minute)*
1	—	3
2	—	4
3	1	2
4	2	5
5	3	4
6	5	8
7	4	2
8	6	4
9	8	6
		(Total time = 38 minutes)

(a diagram that describes the ordering in which work elements should be performed is called a 'precedence diagram')

4. Sketch the precedence diagram and mark the task duration.
5. Assume (for this problem) that the maximum time available at any work station is 10 minutes say. In other words, cycle time is given as 10 minutes. The total duration of all the tasks is 38 minutes which means the minimum number of stations required are 38/10 say 4. The maximum number of stations may be equal to as many as the number of tasks or steps, *i.e.*, nine. It will be tried to achieve the balance with 4 stations.

6. There are two basic concepts of assigning tasks to stations:
   (*a*) *Permutability of tasks.* It means that any number of tasks or steps of a column can be combined to make up their time closer to cycle time, provided their total time does not exceed the cycle time (*i.e.*, time available at a work station). Tasks of even different columns can be combined, provided the precedence constraint is maintained. Analysis is carried out column by column and one can move to next column only after the tasks in the previous column have been assigned to a station.
   (*b*) *Lateral transferability of the tasks.* For making total time of tasks equal to cycle time, tasks or steps may be shifted laterally provided the precedence relationships are maintained.

Using the above two concepts and with a purpose to minimize idle time at a station, Fig. 2.6 is drawn first and shown in 6 work stations. We observe here that the idle time ranges from 0 (at IV) to as high as 6 (at stations III and V) and other stations are also having quite substantial amount of idle time. So, the line is not properly balanced and we attempt to reduce the idle times to the minimum extent at each resultant work station. Thus, Fig. 2.6 is modified as Fig. 2.7 in which, tasks 1, 2 and 3 have been grouped together and assigned station *A*, task 4 has been transferred laterally from column II to column III and has been grouped with task 5, thus occupying station *B*(S-B). Similarly tasks 6 and 7, 8 and 9 have been grouped and placed at stations *C* and *D* respectively. Thus, all the nine tasks have been distributed to four stations now.

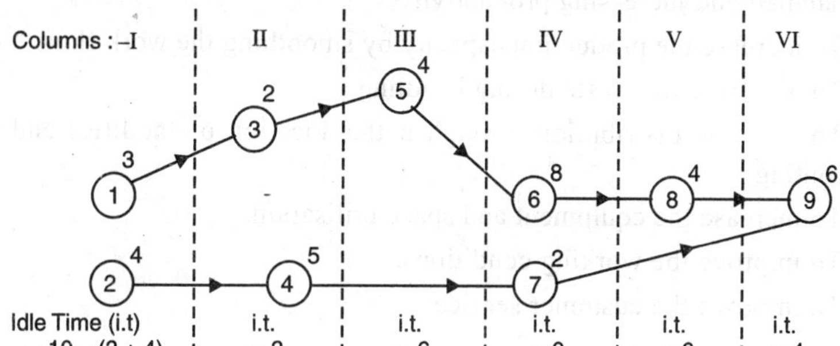

Fig. 2.6. Precedence Diagram

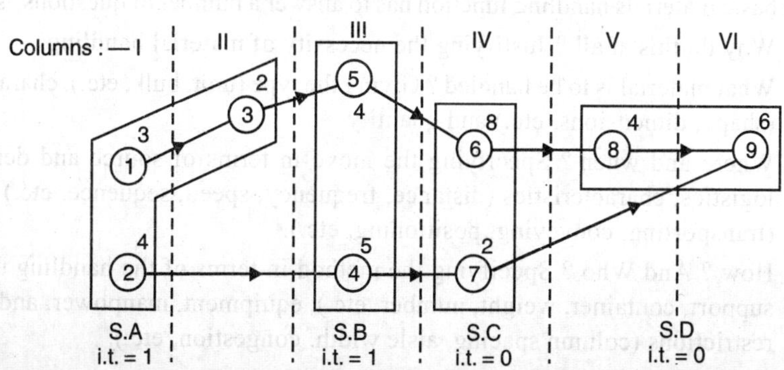

Fig. 2.7. Most Likely Solution (of Line Balancing Problem, Cycle Time is 10 Minutes)

Each of the two stations *A* and *B* are idle for 1 minute (10 – 9 = 1), whereas station *C* and *D* have no idle time (10 – 10 = 0).

2.3. MATERIAL HANDLING

2.3.1. Introduction

We have already discussed the analysis of material flow and the design of layout based on it. We have referred to the selection of material handling equipment and area allocation for it. Materials handling is the art and scinece involving the movement, packaging and storing of substances in any form. Here, we will discuss about the objectives of the material handling system design, basic types of material handling systems and the procedure for the design and selection of material handling systems while developing a plant layout.

Material handling is a system which forms the various factors of movement, transfer, warehousing, in-process handling and shipping into one independent cycle considering the most economical solution for the respective plant.

2.3.2. Objectives and Functions

In order to perform the activities of materials handling the basic goal is to minimise the production costs. This general objective can be further subdivided into specific objectives as follows :

(*i*) To reduce the costs by decreasing inventories, minimising the distance to be handled and increasing productivity.

(*ii*) To increase the production capacity by smoothing the work flow.

(*iii*) To minimise the waste during handling.

(*iv*) To improve distribution through better location of facilities and improved routing.

(*v*) To increase the equipment and space utilisation.

(*vi*) To improve the working conditions.

(*vii*) To improve the customer service.

The analysis of materials handling requirements can be carried out by using travel charts and other quantitative techniques.

The basic materials handling function has to answer a number of questions as follows:

(*i*) Why do this at all ? Justifying the necessity of material handling.

(*ii*) What material is to be handled ? Giving the type (unit, bulk, etc.), characteristics (shape, dimensions, etc.) and quantity.

(*iii*) Where and when ? Specifying the move in terms of source and destination, logistics, characteristics (distance, frequency, speed, sequence, etc.) and type (transporting, conveying, positioning, etc.).

(*iv*) How ? And Who ? Specifying the method in terms of the handling unit (load support, container, weight, number, etc.), equipment, manpower, and physical restrictions (column spacing, aisle width, congestion, etc.).

2.3.3. Basic Materials Handling Systems

The different material handling systems can be classified according to the type of equipment used, material handled, method used or the function performed.

Equipment-Oriented Systems

Depending upon the type of equipment used, there are several systems :

(*i*) Overhead systems
(*ii*) Conveyor systems
(*iii*) Tractor-trailor system
(*iv*) Fork-lift truck and pallet system
(*v*) Industrial truck systems
(*vi*) Underground systems.

Material Oriented Systems

These may be of the following types :

(*i*) Unit handling systems
(*ii*) Bulk handling sysems
(*iii*) Liquid handling systems

A unit load consists of a number of items so arranged that it can be picked up and moved as a single entity such as a box, bale, roll, etc. Such a system is more flexible and requires less investment.

Method Oriented Systems

According to the method of handling and method of production, the material handling systems can be :

(*i*) Manual systems
(*ii*) Mechanised or automated systems
(*iii*) Job-shop handling systems, or
(*iv*) Mass-production handling systems

Function Oriented Systems

The systems can be defined according to the material handling function performed as follows :

(*i*) Transportation systems
(*ii*) Conveying systems
(*iii*) Transferring systems
(*iv*) Elevating systems

2.3.4. Selection and Design of Handling System

The selection and design of the material handling system should be done alongside the development of the layout as each one affects each other. Hence, an integrated approach to the design process is usable. A computerised technique known as COFAD (Computerised Facilities Design) has been developed for integrated handling system and layout design. The steps to be followed in the selection and design of handling systems are as follows :

(*i*) Identification of system
(*ii*) Review of design criteria and objectives of the handling system
(*iii*) Data collection regarding flow pattern and flow requirements
(*iv*) Identification of activity relationships
(*v*) Determining space requirement and establishing material flow pattern

(*vi*) Analysis of material and building characteristics

(*vii*) Preliminary selection of basic handling system and generation of alternatives considering feasibility of mechanisation and equipment capabilities

(*viii*) Evaluation of alternatives with respect to optimal material flow, utilising gravity, minimum cost, flexibility, ease of maintenance, capacity utilisation and other objectives of the system design considering various tangible and intangible factors.

(*ix*) Selection of the best suited, alternative and checking it for compatibility.

(*x*) Specification of the system.

(*xi*) Procurement of the equipment and implementation of the system.

2.3.5. Materials Handling Equipments

After the simplification of the handling method the selection of equipment is important with respect to the different objectives of speed, efficiency-cost, etc. There are both the manual and powered kind of handling equipments. Some of the typical handling equipments are shown in Fig. 2.8. Material handling equipments are broadly classified into four basic types as shown below :

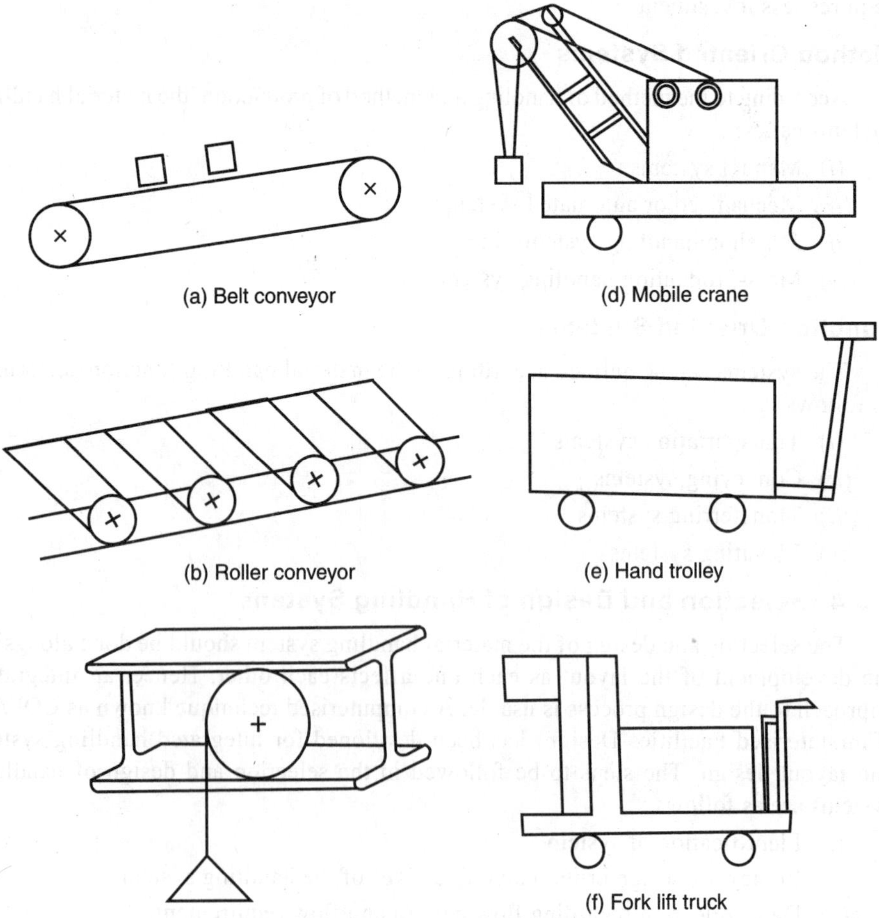

(a) Belt conveyor (d) Mobile crane

(b) Roller conveyor (e) Hand trolley

(c) Mono rail (f) Fork lift truck

Fig. 2.8. Material Handling Equipments

Type I : Conveyers

These are gravity or powered devices commonly used for moving uniform loads from point to point over fixed paths, where the primary function is conveying. Commonly used equipment under this category are :

(*i*) Belt conveyer (*ii*) Roller conveyer (*iii*) Chain conveyer
(*iv*) Bucket conveyer (*v*) Trolley conveyor (*vi*) Screw conveyer
(*vii*) Pipeline conveyer (*viii*) Vibratory conveyer (*ix*) Chute.

Type II : Cranes, Elevators and Hoists

These are overhead devices used for moving varying loads intermittently between points within an area, fixed by the supporting and binding rails, where the primary function is transferring or elevating. Some common examples are :

(*i*) Overhead travelling crane (*ii*) Gantry crane
(*iii*) Jib crane (*iv*) Elevators (*v*) Hoists
(*vi*) Stacker crane (*vii*) Winches (*viii*) Monorail

Type III : Industrial Trucks and Vehicles

These are hand operated or powered vehicles used for movement of uniform or mixed loads intermittently over various paths having suitable running surfaces and clearances, where the primary function is manoeuvering or transporting. These include :

(*i*) Fork lift truck (*ii*) Platform truck
(*iii*) Industrial tractors and trailors (*iv*) Industrial cars
(*v*) Walkie truck (*vi*) Two-wheeled hand truck or trolley
(*vii*) Hand stacker.

Type IV : Auxiliary Equipment

These are devices or attachments used with handling equipment to make their use more effective and versatile. Some common examples are :

(*i*) Ramps (*ii*) Positioners
(*iii*) Pallets and skids (*iv*) Pallet loader and unloader
(*v*) Lift truck attachments (*vi*) Dock boards and levellers
(*vii*) Containers (*viii*) Below the hook devices
(*ix*) Weighing equipment.

2.3.6. Safety Requirements While Using Material Handling Equipments

Millions of industrial accidents occur every year. In these accidents lacs of workers lose their lives every year in the world. Accidents may cause injury which sometimes result in death or permanent total disablement. Accidents in every operation are caused due to over-confidence, carelessness, laziness or variations in performance and the same is true with material handling equipments also. By proper personal contacts between operators and supervisors, accidents can be minimized if not totally eliminated. It is needless to say that safety in a machine is the responsibility of production management and not of safety officer whose function is only advisory. The following five point programme can help in implementing safety requirements while using material handling equipments.

1. Machine Design : The design of the machine should be done in such a way that it should be conductive to safe installation and maintenance of the machine. In both of these aspects of installation and maintenance, danger rises with fixed path equipments such as overhead cranes and powered conveyors, etc. Adequate clearances should be provided for overhead cranes and conveyors for maintenance as well as operational and erection purposes.

For overhead cranes, covering a large area :

(*a*) precautions should be taken to safeguard maintenance personnel who may be invisible from control point as swing clears.

(*b*) facility should be provided to look out starting button and operating levers.

(*c*) cranes should be operated from at least 10 feet away from the high voltage overhead wires.

(*d*) automation on feeding can add to the safety of the worker.

(*e*) metal forming can be made safe by automatic feeding from strip and by automatic ejection of finished component into a pallet.

(*f*) while working with a crane, sudden lifting and panic braking should be avoided.

(*g*) where crane operator cannot see all the parts of load and its path clearly, a signalman should guide the operator,

(*h*) wire ropes should be checked daily for broken strands and excessive wear (to be replaced if found in unworthy condition)

(*i*) the crane operator should always lower the load to the ground before leaving the crane,

(*j*) 'sloshing' or use of hookswing to extend the radius of crane should be strictly prohibited because misjudging the momentum of load always cause accidents.

For fork and platform lifting trucks :

(*a*) the operator's limbs should be kept away from dangerous features such as slides or platforms.

(*b*) battery mountings behind driver should be secured enough to withstand impact of deacceleration and overturning of trucks.

(*c*) a periodic check of brakes, steering apparatus and lift mechanism, etc. should be maintained.

(*d*) the fork lift truck should not be driven over objects lying on the floor.

(*e*) if the truck is without load, the forks should be kept eight inches above the ground.

(*f*) truck routes and crossings must be clearly marked.

(*g*) mirrors should be provided at blind corners and obstacles should be marked in black and yellow diagonals.

For conveyors and elevators :

(*a*) emergency stop buttons or a continuous stop cord should be provided at all danger points.

(*b*) trip switches should be provided to stop equipment beyond a level.

(*c*) trailing rather than leading hooks can reduce the risks of serious accidents.

2. Safety Regulations and Discipline : Accidents can be effectively controlled if not totally eliminated provided safety regulations and discipline are implemented hand in hand to each other. The drivers should be instructed to:

(*a*) face the direction of travel of the equipment or material handling vehicles.
(*b*) never travel with forks raised more than 8 inches above the ground.
(*c*) never exceed the rated load.
(*d*) always put red flags on over-hanging loads.
(*e*) always travel backwards with high and bulky loads.
(*f*) keep the inside of cab tidy. Keep windscreen, mirrors and lights clean.
(*g*) never smoke when refuelling and always stop the engine while refuelling.
(*h*) before working on hydraulic system, make sure that hydraulic pressure is released.
(*i*) never use a naked flame to inspect battery or any leakage.
(*j*) follow traffic rules.
(*k*) never operate unqualified persons to your machine.

3. Training : Training to the operators handling the M.H equipments is the positive element in accident prevention. Proper training can ensure operating the machines at maximum speed which can boost up productivity. The training scheme should be organised by the equipment's manufacturers for their customer drivers. A typical training course for drivers may aim at :

(*a*) technically qualifying him to handle equipment efficiently
(*b*) teaching him to anticipate danger and how to get over it.

4. Safety Incentives : Incentive can prove to be a booster dose to the performance of the drivers. A driver may be allocated 1000 points at the beginning of the year, which may be deducted for offences such as—not following safety regulations, non-reporting of equipments' defects and causing accidents, etc. A suitable reward can ensure reduction of accidents. Displaying prominently on notice board the statistics of monthly accidents (or any periodic frequency) can cause safety awareness among the workers and drivers.

2.3.7. Cost of Material Handling Equipment and Handling Costs

After selecting from a great variety of material handling devices, on the basis of technical factors, they are compared from the engineering and economic points of view. In economic evaluation of various types of devices both the total "capacity outlay" and the "operational handling costs" should be taken into consideration. The capital outlay includes the cost of equipment, the cost of its erection, transporting and construction costs in installation and operation. Operational handling costs include :

(*a*) Wages and salary of the personnel plus additions of social services.
(*b*) Cost of electric power
(*c*) Cost of lubricating, wiping, cleaning and consumables there to.
(*d*) Cost of repair and maintenance.

Moreover, account should also be taken of losses due to depreciation of the equipment during handling. The expenditure on the maintenance of the material handling installation, annual deductions for the amortisation of mechanical devices and the fund assigned for their general overhead should also be taken into account.

MODEL QUESTIONS

1. (*a*)What are the various factors affecting plant location ?
   (*b*)What factors would influence the choice of multistoreyed buildings for manufacturing industry ? Give a list of industries located in India for which such buildings are preferable or essential.
   [**Ans.** Flour mills, bakery, printing industry, chemical industry, etc., are some of the industries for which multistoreyed buildings are suitable. In flour mills, raw materials are lifted to the topmost floor by conveyers and work-in-process is automatically transferred to the successive floors are processes are completed. In a bakery the processes are usually commenced on the ground floor where baking is completed. The work-in-process is transferred automatically by conveyor belts to the first floor where varieties are packed. In machine composing in printing industry, two floors one above the other are used for completing the printing process. In chemical industry, liquid and powdered products may be transported by the force of gravity.]
2. Make an outline of the different factors to be taken into account in the preliminary estimation of the total floor area needed for the layout of a factory for a manufacturing concern.
   [**Ans.** Total floor area is the sum of the estimated floor areas for : (1) aisles, (2) column, (3) material handling equipment, (4) production department which includes work station, storage space for work in process, aisles and clearance area, (5) service departments which include receiving and shipping departments and space for raw materials for finished goods stores. Storage space for regular stores materials is calculated with reference to minimum stores levels. Elaborate on the above points.]
3. What is plant layout ? Discuss the various objectives of a good plant layout.
4. (*a*) When does a location decision arise ?
   (*b*) What are the advantages derived from an efficient plant layout ?
5. Enumerate the basic types of plant layout and write their characteristic features in brief.
6. Outline the features of :
   (*i*) indiscriminate layout
   (*ii*) grouped layout
   (*iii*) continuous layout for machine tools in an engineering works.
   [**Ans.** Indiscriminate layout is the combination of product layout and process layout. Grouped layout stands for process layout and continuous layout is product layout.]
7. Name at least 3 different industries in India in which continuous layout of machine is adopted.
   [**Ans.** Petroleum refining, automotive industry and chemical industry]
8. Discuss what do you consider to be a good practice in the layout of shops and equipments for different classes of engineering works?
9. Discuss how plant layout and material handling methods depend on the nature of product and the quality of manufacture? Give examples.
10. Distinguish fully between the product layout and process layout in shop.
11. What is the difference between functional and product layout of machines.
12. Explain the importance of travel chart in effective layout of a production plant. Prepare a travel chart for a hypothetical engineering concern with 5 functional deptts, *i.e.*, foundary, forging, machining, welding and inspection. Given this travel chart proceed to find the locations of different deptts.
13. Discuss the different principles of plant layout in industry.

14. What are the symptoms of a poor plant layout ?
15. Enumerate and explain briefly the factors which affect the layout in an organization.
16. How does a cellular layout differ from a process layout ?
17. What are the conditions which must exist to make practicable a layout of machines for contineous production ? Name some of the difficulties inherent in the layout and how they may be removed.
18. What are the different kinds of material handling equipment used? Explain briefly.
19. Describe the concept of Line Balancing in short.
20. The best handling is "no handling" and if this cannot be achieved "minimum handling" should be tried. Justify the statement.

❑❑❑

CHAPTER 3

SALES FORECASTING

INSIDE THIS CHAPTER

3.1. INTRODUCTION

Business action takes today must be based on yesterday's plan and tomorrow's expectations. Plans for the future cannot be made without forecasting events and their relationships with other events. Sales forecasting is the most important activity in a concern because all other activities depend on it. Forecasting is a systematic attempt to probe the future by inference from known source. Every firm is keen to know the expected demand for its products, how much of a given product it could sell in a given time, whether the sales would increase or decrease from the current levels and by how much and what would be the share of the market it can secure during the specified period. This knowledge is required by the firm for its survival and growth. It forms the basis for the requirements of the raw materials, equipment, jigs and fixtures, labour force and capital, etc. All these decisions are basically related to the size of production which in turn can be determined from potential demand for the product. Thus, the vital point of all decisions related to production strategy is the sales forecast for a specified period.

3.2. DEFINITION

According to American Marketing Association : **"Sales forecasting is an estimate of sales in monetary value or physical units for a specified future period under a proposed marketing plan or programme and under an assumed set of economic and other forces outside the unit for which the forecast is made."**

Sales forecast helps to predict the future taking note of the past and being aware of the present.

3.3. FACTORS AFFECTING SALES

Following are the factors that affect the sales of a company :

(*i*) Seasonality of the business.

(*ii*) Relative state of economy.

(*iii*) Diect and indirect competition.

(*iv*) Political events.
(*v*) Styles or fashions.
(*vi*) Consumer earnings (purchasing power).
(*vii*) Population changes.
(*viii*) Weather.
(*ix*) Productivity changes.

3.4. IMPORTANCE OF SALES FORECASTING

The data of sales forecasting is very essential for an industrial enterprise or a business firm. In producing units, the importance of sales forecasting is greater because of their varied problems of producing and distributing the products on a large scale. Sales forecasting is a guiding factor for a firm because it enables the firm to produce the actual quantity of the required quality product, at the right time and at a reasonable price. The firm can arrange the various resources, such as raw materials, equipments, machines, accessories, labour and buildings, etc. required for production well in advance. Only a few firms produce products on the basis of the orders received from customers; most of the firms produce in 'anticipation of demand'. Forecasting helps a firm to assess the probable demand for its products and plan it accordingly. With the help of sales forecasting, the producer can prepare production schedule and can reduce inventory, production and other related costs.

Sales forecasting is the basis for planning various activities, such as production activities, price policies, marketing planning, programmes policies and strategics. Thus, overall planning can be done only on the basis of sales forecasting.

Sales forecasting helps management to take correct decisions well in time and thus, helps avoid unnecessary hardships to the firm. In this way, it also helps in reducing the wastage of men, materials and machines, etc. Thus, it measures the correctness of management decisions.

Sales forecasting is valuable because it prepares the management for change. Sudden and temporary changes can be avoided to diffuse the situation. It is possible only with the help of sales forecasting data. Sales forecasting also sets the goal for the sales department. If the sales department accomplishes the goal set for it, it can be regarded as an efficient and well-organized department. In this way, sales forecasting helps in measuring the efficiency of the sales department.

The importance of sales forecasting lies in its correctness. The above mentioned advantages of sales forecasting can be taken only if the forecasting is accurate and close to the actual performance. For this purpose, a review of estimates should be made from time to time.

According to John D. Luth, "A good sales forecast is undoubtedly the most important single planning tool. It is the basis of a sound planning. All budgeting starts with the sales forecast. Financial planning for working capital requirements, plants expansion and other needs are based on anticipated sales. Scheduling of all production resources and facilities such as manpower needs, raw material requirements and determining the rate of production output depends upon the sales forecast."

3.5. OBJECTIVES OF FORECASTING

3.5.1. Short Term Objectives of Forecasting

Depending on the nature of work (project) the time period may vary from 2 to 12 months or so.

(i) *Formulation of suitable production policy :* Forecast helps to formulate a suitable production policy so that problems of under-production or over-production may not arise.

(ii) *Regulate supply of raw material :* It is possible to evaluate the requirements of raw materials in future so as to ensure regular and continuous supply of materials on the basis of estimated sales and also to control the size of inventory at economic level.

(iii) *Best utilization of machines :* The operations can be so planned that the machines are utilized to their maximum capacity.

(iv) *Regular availability of labour :* One of the objectives of sales forecasting is also to arrange for trained personnel and non-technical workers so that they might not experience any shortage of personnel and at the same time they don't remain idle when the production is curtailed.

(v) *Price policy formulation :* Sales forecasts enable the management to formulate some appropriate pricing mechanism, so that the level of prices does not fluctuate too much in the periods of depression or inflation.

(vi) *Forecasting of short term financial requirements :* On the basis of sales forecast it is possible to determine the financial requirements of the enterprise for the production of desired output; and arrange it accordingly much in advance.

(vii) *Setting the sales target:* Sales forecasts are calculated for different market segments and then the sales targets for various territories are fixed accordingly. This later on becomes the basis to evaluate and control sales performance.

3.5.2. Medium-term Objectives of Forecasting (Time Horizon is Approximately 4 Months to 24 Months)

It is popularly called sales and operations planning in practice and relates to aggregate planning. To handle seasonality in demand, the forecast (medium) is used to build up seasonal inventory budgeting and workforce planning. It allows a firm to have a horizon of 4-24 months for managing resource planning and budgeting and forecast is carried out at aggregate levels (across product variants at category level). Both time-series models and causal models are used for medium term forecasting. Causal models are prepared as they can handle the turning points (growth to maturity and maturity to decline) in the life cycle better.

3.5.3. Long-term Objectives of Forecasting

Time period may range from 5 to 10 years or even longer for ship building, petroleum refinery and electricity generation.

(i) *Deciding Plant Capacity*: The long run objective of sales forecasting is to plan plant capacity in accordance with the demand. The size of the plant can be determined such that the output conforms with sales requirements. Too small or too large size of the plant may not be in the economic interest of the enterprise.

By studying the demand pattern for the product and the forecasts for future ; the enterprise can plan for a plant with output of desired capacity.

(ii) *Manpower Planning :* Reliable and accurate forecast can help the management to assess the appropriate manpower requirements. This can ensure best manpower facility to carry out the production in the long run without any hindrances.

(iii) *Estimating Cash Inflows :* Cash inflows from sales can well be estimated through sales forecasting by determining the cash and credit sale ratio. It can also help to plan for credit policy of the firm.

(iv) *Determining Dividend Policy :* The profits can also be forecast on the basis of gross profit ratio on sale and dividend policy can be determined.

(v) *Planning of Long-run Production :* Long-run production planning also depends upon sales forecasting. In the long-run product has to be adjusted to the market demand and other conditions.

(vi) *Long-run Financial Requirements :* Sales forecasting also helps to determine the long-run financial requirements of the organisation for working capital as well as for capital expenditure.

(vii) *Budgetary Control over Expenditure :* In forecasting the sales, all the activities are to be forecast and for this purpose budget is to be prepared for the income and expenditure of the organisation. The budgeting figures for income and expediture may then be compared with the actual performance and any variation is removed. Thus, budgetary control over expenditure becomes possible.

3.6. PROCESS OF SALES FORECASTING (STEPS IN SALES FORECASTING)

The following are the main steps in demand forecasting :

1. **Determining the objective of forecast :** Certain points in this respect should be very clear before taking up the forecasting test such as period of forecasting (short term, long term); area of sales forecasting, unit of sales forecasting (*i.e.* in quantities or values), the time, labour and money to be employed on forecasting. All these points are determined taking into account the objectives of sales forecasting.
2. **Sub-divide the task of forecasting :** Sub-divide the forecasting programme into homogeneous groups according to product, area, activities or customers. The total sales forecast of the company will be the sum total of all the groups.
3. **Determine the relative importance of factors :** So that due weightage may be given to different factors affecting forecast.
4. **Select the method to be used for forecasting :** The method is to be selected by the appropriate authority taking into account all the relevant situations, purpose of forecasting and the degree of accuracy.
5. **Collect and analyse the data :** By applying the appropriate method, the necessary data for forecast are collected, tabulated and cross-checked. The data are interpreted by using statistical techniques. It may be called as the preliminary sales forecast which forms the basis for final sales forecast.
6. **Study the correlation between sales forecast and sales promoting plans :** Making the forecast reliable, the sales promotion plans such as advertising

policy, personal selling and other policies should be reviewed with reference to the preliminary sales forecast.

7. **Study of competitors' activities :** Volume of sales of a company is largely affected by the activities of competitors and, therefore, it is essential to study the competitor's activities, policies, programmes and strategies and also their effects on the market and adjust the forecasting accordingly.
8. **Prepare final sales forecasts :** The preliminary sales forecast results are converted into final sales forecast relating to the products and territories involved. The aggregate of sales forecasts of different products, or territories or customers or activities may form the sales forecasts of the enterprise.
9. **Evaluation and adjustments :** The actual sales performance in the coming period should be reviewed and evaluated from time to time. The evaluation may be made monthly, quarterly, half yearly or yearly. The forecast figures may be revised in the light of difficulties experienced during the course of actual operations. On the expiry of forecasts period the actual sales and forecast sales should be compared and causes of variation found out which may help to improve the next period sales forecasts.

3.7. FORECASTING MODELS (TECHNIQUES OF FORECASTING)

The important methods of sales forecasting are the following :

3.7.1. Qualitative Techniques

Judgement based techniques when hard data are scarce or difficult to have availability. These are as follows :

1. Survey of Buyer's intention or the user's expectation technique. (Customer survey)
2. Collective opinion or the sales force composite technique. (Sales force survey)
3. Group executive judgement or executive judgement technique. (Executive committee consensus)
4. Experts' opinion technique. (Educated guess)
5. Market research
6. Delphi method
7. Historical analogy

1. Survey of Buyer's Intentions or the User's Expectation Technique (Customer Survey). In the technique, the actual users of the product are contacted directly and asked about their intention to buy the company's products in an expected time, usually one year. Total sales forecasts of the product are then estimated on the basis of advice and willingness of various customers. This is more direct method of sales forecasting.

Advantages

Sales forecast is based on information received or collected from the actual users and so, the estimates are correct.

(*i*) It provides a trend feel of the market and of the thinking behind the buying intentions of the actual users. This helps the organization to develop a new product easily.

(*ii*) This technique of sales forecasting is more appropriate where total numbers of users of the product are determined and a new product is to be introduced, for which no previous records are available.

(*iii*) It is most suitable for short-run forecasting.

Disadvantages

The disadvantages of this technique are :

(*i*) This technique is not practically possible in case of numerous customers such as in case of household goods.

(*ii*) This technique is based on the intentions and expectations of the customers, which are subject to change.

(*iii*) It is not appropriate for long-run sales forecasting.

(*iv*) This technique is passive and does not depict or measure the variables that are under management control.

2. Collective Opinion or the Sales Force Composite Technique (Sales Force Survey) : In this technique, views of salesmen, branch managers, area managers and sales managers are secured for the different segments of the market. The salesmen, being close to actual users, are required to estimate the expected sales in their respective territories and sections. The estimates of individual salesmen are then consolidated to find out the total estimated sales for a specific period. These estimates are then further examined by successive executive levels of the organization in the light of various factors like proposed changes in product design, advertising, selling prices and competition, etc., before they are finally merged for forecasting.

The **advantages** of this technique are :

(*i*) The predictions are more accurate and reliable as the forecasts are based on the knowledge of salesmen and others connected with sales.

(*ii*) This technique is simple and inexpensive and useful for forecasting of sales for a new product.

(*iii*) It increases the confidence and motivates the sales people.

(*iv*) It places the responsibility of sales forecast on those who are expected to produce results.

The **disadvantages** of this technique are :

(*i*) The accuracy of sales forecast figures is always doubtful because it is possible that the salesmen or executives have made wrong estimates due to their optimistic or pessimistic nature.

(*ii*) This technique is not suitable for long-term production planning because the salesmen are not aware of the broad economic patterns and therefore, cannot forecast for long term.

(*iii*) If the salesmen are aware of the fact that their sales quota will be fixed on the basis of this information, they may intentionally under-estimate the sales.

(*iv*) It is a subjective approach.

3. Group Executive Judgement or Executive Judgement Technique (Executive Committee Consensus) : This is the process of combining, averaging or evaluating the opinions and views of top executives in some other way. In this method, the opinions of the executives of different fields such as marketing, finance productions, etc. are gathered to make the forecast.

The main **advantages** of this technique are :

(*i*) This technique is easy and simple. Forecasts can be done by the executives without delay.

(*ii*) There is no need for collecting elaborate statistics for the forecast because forecasts are based on the knowledge and experience of the experts of different departments.

(*iii*) It permits policy of specialized viewpoint.

(*iv*) It is the only feasible means of forecasting in the absence of sufficient data.

The **disadvantages** of this method are :

(*i*) It has no factual basis for estimation, hence it is an inferior technique.

(*ii*) It consumes costly executive time.

(*iii*) Accuracy cannot be claimed under this technique because accuracy depends upon the ability, knowledge and efficiency of the executives.

4. Experts' Opinions Technique (Educated Guess) : In this technique, the organization collects opinions from specialists in the field outside the organization. These experts' opinions are collected from the newspapers, wholesalers and distributors for the company's products, and professional experts. By analysing these opinions, deductions are made for the company's sale forecasts by the experts.

The **advantages** of this technique are :

(*i*) Forecasts can be done easily and speedily.

(*ii*) Different experts' ideas and opinions are taken into consideration for forecasting.

(*iii*) It is suitable where past records of sales are not accurate.

(*iv*) This technique is economical because the organization does not have to collect data. The expenses for seeking the opinions of experts are much less than the expenses of an actual survey.

The **disadvantages** of this technique are :

(*i*) Market segmentation on the basis of customers is not possible.

(*ii*) The reliability of forecasting data is always doubted because forecasting is based on opinions and suggestions and not on the facts.

5. Market Research : In this technique, the salesmen sell the product in a part of the market for some time and evaluate the sales for the full market on the basis of the results of the test sales. This technique is appropriate when the product is quite new in the market or good estimators are not available or when the buyers have not prepared their purchase plan.

The **advantages** of this technique are :

(*i*) This method is most suitable for introducing a new product in the market.

(*ii*) Any defect in the product comes to light and can be removed immediately to make the product successful in the market.

(*iii*) Sales forecasts are more reliable because they are based on actual results.

The **disadvantages** of this technique are :

(*i*) It takes a long time to test the market.

(*ii*) The sales forecasts data are planned on the results of just one part of the market. As all the parts of the market are not homogeneous, these forecasts may not be reliable.

6. Delphi Method : It is based on the personal judgment of a panel of experts drawn from within and outside the organisation. But it is different from conferences and panel discussions where direct communication between participants leads to a clash of personal egos and group interactions. In Delphi method, an attempt is made to eliminate these undesirable effects so as to arrive at a consensus rather than a compromise.

7. Historical Analogy : Typically all products go through a life cycle of introduction, growth, maturity and decline. Based on the experience of similar products in the past, one can make use of the diffusion model with the overall market size and can estimate demand distribution over a products life cycle. Relevent parameter can be estimated from the life cycle demand data available for products of similar characterstics introduced in the past.

3.7.2. Quantitative Techniques (Analytical and Statistical Methods)

Basically all analytical and statistical methods of forecasting project historical information into the future. These are based on the assumption that future patterns tend to be the extention of past ones, and that one can make useful predictions by studying the past behaviour *i.e.*, the factors which were responsible in the past will also be normally operative to some extent in the future.

The methods used when enough hard data exist to specify relationship between variables. These are as follows :

1. Moving average method and weighted moving average method
2. Exponential smoothing method
3. Regression analysis—Least square (Trend projection method)

These are explained further.

3.7.3. Measures of Forecast Accuracy (Errors Estimates in Forecasting)

Demand forecast influences most of the decisions in all the functions. Hence, it must be estimated with the highest level of precision. Some common measures are inevitable to measure the accuracy of a forecasting technique. The measure may be an aggregate error (deviation) of the forecast values from the actual demands. The different types of errors which are generally computed are as presented below :

1. Mean Absolute Deviation (MAD).
2. Mean Square Error (MSE) and Standard Error of Estimate
3. Mean Forecast Error (MFE)
4. Mean Absolute Percentage Error (MAPE)

The formula for error is given below

$$e_t = D_t - F_t$$

where, D_t = demand for the period t,
F_t = forecast demand for the period t, and
e_t = forecast error for the period.

1. Mean Absolute Deviation (MAD) : It is the mean of absolute deviations of forecast demands from actual demand values. The MAD is sometimes called as the mean absolute error (MAE).

$$\text{MAD} = \frac{\sum_{t=1}^{n} |D_t - F_t|}{n}$$

where, D_t = Actual demand for the period t
F_t = Forecast demand for the period t
n = Number of time period used

2. Mean Square Error (MSE) : Mean square error is the mean of the squares of the deviations of the forecast demands from the actual demand values. Usually the effects on operations of small errors are not serious. These errors may be smoothed out by inventory or overtime work. It will be difficult to have smoothed values for forecast if there are few large errors. Consequently, a method of measuring errors that penalizes large errors more than small errors is some time desired. The mean square error (MSE) provides this type of measure of forecast error.

$$\text{MSE} = \frac{\sum_{t=1}^{n} (D_t - F_t)^2}{n}$$

where, D_t = Actual demand for the period t
F_t = Forecast demand for the period t
n = Total number of periods used.

Standard Error of Estimate : It is the square root of MSE.

3. Mean Forecast Error (MFE) : Mean forecast error (MFE) is the mean of the deviations of the forecast demands from the actual demands.

$$\text{MFE} = \frac{\sum_{i=1}^{n} (D_t - F_t)}{n}$$

where, D_t = Actual demand for the period t
F_t = Forecast demand for the period t
n = Number of years (time period) used.

4. Mean Absolute Percentage Error (MAPE) : Mean absolute percentage error (MAPE) is the mean of the per cent deviations of the forecast demands from the actual demands.

$$\text{MAPE} = \frac{1}{n}\sum_{i=1}^{n} \frac{|D_t - F_t)|}{D_t} \times 100$$

Suppose that a forecast of 165 units had been made for the demand in every period for the data given in Table 3.1. The calculations of these errors are as shown in the last two columns of Table 3.1.

Table 3.1. Calculation of Various Errors

T	Demand D_t	Forecast F_t	Deviation $D_t - F_t$	Absolute Deviation $\|D_t - F_t\|$	Squared Error $(D_t - F_t)^2$	Percentage Error $\frac{D_t - F_t}{D_t} \times 100$	Absolute % Error $\frac{\|D_t - F_t\|}{\|D_t\|} \times 100$
1	150	165	–15	15	225	–10.000	10.000
2	160	165	–5	5	25	–3,125	3.125
3	165	165	0	0	0	0.000	0.000
4	175	165	+10	10	100	5.710	5.710
5	180	165	+15	15	225	8.330	8.330
Σ			+5	45	575		27.165

$$\therefore \quad \text{MAD} = \frac{45}{5} = 9, \ \text{MSE} = \frac{575}{5} = 115, \text{MAPE} = \frac{27.165}{5} = 5.433\%$$

and $\text{MFE} = \frac{+5}{5} = +1.$

Standard Error of Estimate = $\sqrt{\text{MSE}} = \sqrt{115} = 10.72$

3.7.3.1. Moving Average Method and Weighted Moving Average Method

A simple moving average is an arithmetic average of a set of observed values from the present time period to a certain time period to the past and then using that average as the forecast for the time period in the immediate future. An appropriate period (usually 3 or 5 or 7 years is taken in practice, as the period in even numbers will create a problem in centralizing the values of the averages) is to be selected for which the moving averages are to be calculated and the calculated values of the moving averages present the basis for determining the expected amount of sales. The example shown below will explain the application.

Example 3.1. The following series relates to the annual sales in thousands of a product during the period 1975 to 1982. Find the trend of the sales using 3 yearly moving averages.

Solution:

Year	Sales in thousands (₹)	3-yearly moving total	3 yearly moving average (trend values)
1975	16		
1976	18	16 + 18 + 15 = 49	$\frac{49}{3} = 16.33$
1977	15	50	16.67
1978	17	52	17.33
1979	20	59	19.66
1980	22	67	22.33
1981	25	71	23.66
1982	24		

Weighted Moving Average Method : Equal weightage (importance) is given to the all previous years in case of simple moving average method. But sometimes it becomes necessary to attach more importance (weightage) to some value compared to others depending on the condition of the situation. Normally for forecasting sales data for future, more weightage are given to the immediate past year (period) compared to earlier years. (The sum of the weights = 1). The example given below will explain the application.

Example 3.2. Compute a weighted three years moving average for 1983 where the weights are 0.5 for the latest previous year and 0.3 and 0.2 for the earlier two years in that order respectively. (Refer Exp. 3.1)

Solution: A 3-years weighted moving average forecast for 1983.

$$= \frac{\text{Demand } 1982 \times 0.5 + D_{1981} \times 0.3 + D_{1980} \times 0.2}{0.5 + 0.3 + 0.2}$$

$$= \frac{24 \times 0.5 + 25 \times 0.3 + 22 \times 0.2}{1}$$

$$= 12 + 7.50 + 4.40$$

$$= ₹\ 23.90 \text{ thousand}$$

Advantages (MA Method)

(*i*) The moving average method provides a simple and good estimate. In this method, equal weightage is assigned to all the periods chosen for averaging.

(*ii*) The process of averaging reduces influence of the fluctuations.

Disadvantages

(*i*) Records of the demand data have to be retained over a fairly long period.

(*ii*) If demand series depicts trends as against the stationary level, the moving average method would provide forecasts that lags the original series.

(*iii*) Choice of period of moving average is difficult.

(*iv*) It cannot be applied if some observations are missing.

3.7.3.2. Exponential Smoothing Method

An improvement over the moving average method, exponential smoothing method tries to overcome the limitations of the moving average method and eliminates the necessity of keeping extensive records of past data. It represents a weightage average of the past observations. The most recent observation is assigned the highest weightage and it decreases in geometric progression as we move towards the older observations. This method is particularly useful when forecasts of a large number of items are made. This method is also adaptable for trend correction and smoothing of forecast errors. It is one of the most accurate statistical techniques for forecasting, since the most recent observations which are likely to reflect more upto date average of the series are given more weightage.

The fundamental concept of exponential smoothing method is that

$$\text{New forecast} = \text{Old forecast} + \alpha\,(\text{Latest observation} - \text{Old forecast})$$

$$\text{or } F_t = F_{t-1} + \alpha(D_t - F_{t-1})$$

where, α = Smoothing coefficient whose value lies between 0-1

The actual value of α determines the balance between sensitivity and stability and it has been found appropriate to have α between 0.1 to 0.5. Use low value of α if series is stable and high vale when the series is fluctuating (If cyclic fluctuations are predominant in forecast, then low value of α should be selected and if the long term fluctuations are more dominant, then select high value of α).

Example 3.3. A firm uses simple exponential smoothing with α = 0.2 to forecast demand. The forecast for the first week of January was 400 units, whereas actual demand turned out to be 450 units.

(*i*) Forecast the demand for the second week of January.

(*ii*) Assuming that the actual demand during the second week of January turned out to be 460 units, forecast the demand upto February third week taking the subsequent demands as 465, 434, 420, 498 and 462 units.

Solution. (*a*) Forecast for the 2nd week of January (made in the 1st week of January)

$$= F_t = F_{t-1} + \alpha(D_t - F_{t-1})$$

$$= 400 + 0.2\ (450 - 400) = 410 \text{ units}$$

Forecast for the 3rd week of January = 410 + 0.2 (460 – 410) = 420 units

Forecast for the 4th week of January = 420 + 0.2 (465 – 420) = 429

Forecast for the 1st week of February = 429 + 0.2 (434 – 429) = 430

Forecast for the 2nd week of February = 428

Forecast for the 3rd week of February = 442

3.7.3.3. Regression Analysis – Least Squares Method (Trend Projection Method)

In this method, a line, $y = a + bx$ is identified which fits the data best, by examining the sum of their squared deviations, and the line with the smallest sum of square errors is determined. The least square line is determined using data from '*n*' periods and unknown parameters *a* and *b*. This is done by solving a set of two linear equations:

$$\Sigma y = na + b\Sigma x$$

$$\text{and } \Sigma xy = a\Sigma x + b\Sigma x^2$$

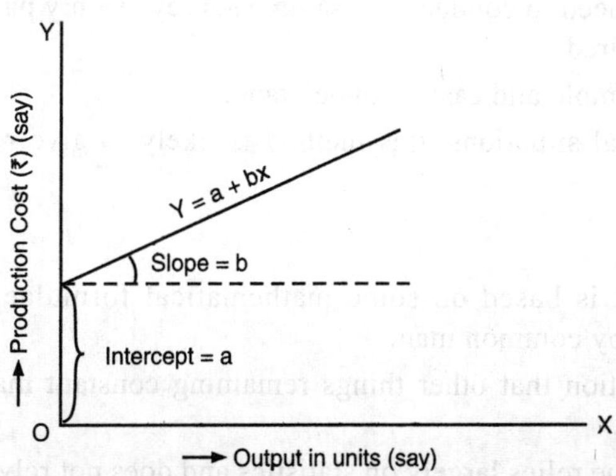

Fig. 3.1

The following examples will illustrate the application clearly.

Example 3.4. The annual sales of a company are as given below :

Year :	**1980**	**1981**	**1982**	**1983**	**1984**
Sales :	**50,000**	**65,000**	**75,000**	**52,000**	**72,000**

By the method of least square, find the trend values for each of the five years and also predict the sales for 1985.

Solution.

Year (X)	*Sales (Y) (in 1000 ₹)*	*Deviation of X from 1982 (x)*	x^2	*xy*	*Trend value* $y_c = a + bx$
1980	50	–2	4	–100	56.60
1981	65	–1	1	–65	59.70
1982	75	0	0	0	62.80
1983	52	1	1	52	65.90
1984	72	2	4	144	69.00
$n = 5$	$\Sigma y = 314$	$\Sigma x = 0$	$\Sigma x^2 = 10$	$\Sigma xy = 31$	

The equation of the line is $y = a + bx$, then $a = \frac{\Sigma y}{n} = \frac{314}{5} = 62.80$

and $b = \frac{\Sigma xy}{\Sigma x^2} = \frac{31}{10} = 3.1$

$\therefore$ Required equation is $y = 3.10 + 62.80x$

Trend values for 1980 is $y_{1980} = 3.10 + 62.80 \times (-2) = 56.60.$

Similarly, the trend values for other years can be computed.

Sales for 1985 = $y_{1985} = 3.10 + 62.80 \times (3)$

= ₹ 66.10 thousand (Note : The number of years is ODD)

If the number of years is **EVEN** say 8, then the chosen values of x would be –7, –5, –3, –1, 1, 3, 5, 7 and rest of the procedure would be the same for the calculation of trend values.

Advantages

(*i*) There is no need to conduct any sample survey as only past information about sales is required.

(*ii*) Method is simple and easy to understand.

(*iii*) Under normal situations, this method is likely to give reliable and accurate results.

Disadvantages

(*i*) The method is based on some mathematical formulae which may not be understood by common man.

(*ii*) The assumption that other things remaining constant may not hold good in practice.

(*iii*) The technique relies largely on statistics and does not rely much on individual skill and experience. Moreover, if requires technical skill and experience for it to be effective.

3.8. CONCEPT OF TIME SERIES IN FORECASTING

The time series analysis consists of determining the trend underlying the demand and extrapolate the future trend from it. Statistical methods are used to determine the trend. Time series consists of the following four components:

(*a*) **Trend :** The long period tendency of the data to increase or decrease is referred to as a secular trend. It is a long term historical component of the time series which indicates overall growth or decline of the business over time.

(*b*) **Cyclical Fluctuations:** These are the oscillating patterns of increase or decrease around the trend. Cyclical fluctuations account for some of the variance between the trend line and the data points. The magnitude, timing and pattern of cyclic fluctuations vary much widely and the awareness of the effect of cyclic variation should influence the forecaster.

(*c*) **Seasonal Variations:** These occur with some degree of regularity in a span of one year and are annually repetitive. These variations are caused by climate conditions, social customs and festivals, etc. For example, the sale of paints is the highest during Diwali, Sale of umbrellas and rain-coats during monsoon time and warm clothes during winter season.

(*d*) **Irregular Variations:** These occur without any fixed pattern . These are chance variations that are not explained by trend, cyclic or seasonal variations. These may account for natural calamities like floods, famines, earthquakes and epidemics, etc.

3.9. ADVANTAGES AND LIMITATIONS OF SALES FORECASTING

Advantages of Forecasting

(*i*) *Helps in Effective Planning :* Forecasting helps in effective planning by providing a scientific and reliable basis for anticipating future operations such as sales, production, inventory, supply of capital, etc.

(*ii*) *Helps in Removing the Weaknesses of Organisation Structure :* Since it leads to rationalisation of various procedures for the achievement of organisational objectives.

(*iii*) *Helps in Better Co-ordination :* It helps in better co-ordination of various resources which leads to better utilization of resources and reduction in waste and inefficiencies.

(*iv*) *Achieves Co-operation in the Enterprises :* Various executives at different levels participate in the process of forecasting. This creates a sense of belongingness among them.

(*v*) *Provides a Basis for Effective Control :* Forecasting provides a basis for effective control by providing information where higher degree of control is to be exercised. The managers can predict the weaknesses of their departments through forecasting.

(*vi*) *Important at the National Level :* Economic forecasts of various factors at the national level help in planning for economic development.

Limitations of Forecasting

Forecasts are only estimates of future condition. They can never be actual position. They can only give the best estimate of future course of events, but they can never be hundred per cent accurate and reliable. The following are some of the limitations of sales forecast :

(*i*) Forecasting is based on postulations and assumptions and hence, it is subject to some guess work and possibility of error.

(*ii*) Forecasting is usually based on past data but future may not be a copy of the past.

(*iii*) Changes in consumer's need, taste, fashion and style, etc. may cause inaccuracy in forecast.

(*iv*) There may be lack of history in case of a new product.

(*v*) Forecasts are not full-proof and condition-proof and if there are changes in the general economy of the country ; they may not materialise.

(*vi*) Development of new products, materials and methods may introduce error in the sales forecast of a particular product.

(*vii*) There may be lack of efficient and experienced sales force.

(*viii*) Short term forecasting is more accurate than long term forecasting and hence, its usefulness is limited to short-term purposes.

MODEL QUESTIONS

1. Define "Sales Forecasting", explain its importance.
2. State and explain the objectives of:
   (*i*) Short term forecasting (*ii*) Long term forecasting
3. What are the advantages and limitations of sales forecasting ? Explain briefly.
4. Describe the various steps involved in the sales forecasting in short.
5. What are the various factors affecting sales of an organization ? Discuss each of these briefly.
6. Describe the following methods of sales forecasting:
   (*i*) Moving average method
   (*ii*) Survey of buyers intention method.
7. Describe "Least Squares Method" of sales forecasting with its merits and demerits.
8. Are the Moving Average Method and the exponential smoothing method of forecasting related ? How and why ?
9. Find the trend line by least square method for the data below :

Year :	1975	1976	1977	1978	1979	1980	1981
Demand in 1000 units :	85	75	80	72	65	60	55

Also estimate the demand for the year 1984.

[**Ans.** $y = 70.285 - 4.822x$, 40.521 thou. units]

10. Fit a straight line trend equation by the method of least squares and estimate the trend values and also get the value for the year 1969.

Year	1961	1962	1963	1964	1965	1966	1967	1968
Value	80	90	82	83	94	99	92	104

[**Hint :** Take the years as –7, –5, –3, –1, 1, 3, 5, 7, $y = 91.75 + 1.25x$ and $y_{1969} = 103$ units]

11. An old forecast of 39 is given initially and the smoothing coefficient is 0.10. Smooth exponentially the following observations :
    41, 41, 34, 39, 36, 35, 40, 36, 41 and 33.

[**Ans.** 39.20, 39.38, 38,84, 38.86, 38.57, 38.22, 38.39, 38.16, 38.44, 37.90]

12. What do you understand by exponential smoothing ? What do the following values of smoothing coefficient indicate ? $\alpha = 0$, $\alpha = 1$, α = low value, α = high value.

❑❑❑

CHAPTER 4

MATERIALS MANAGEMENT

INSIDE THIS CHAPTER

❐ Inventory Control (Meaning and Definitions); ❐ Objectives of Inventory Control; ❐ Quantity Standards; ❐ Economic Order Quantity (EOQ); ❐ Selective Inventory Management; ❐ Purchasing; ❐ Procedures, Forms, Records and Reports; ❐ Stores Accounting and Verification Systems; ❐ Modern Materials Management Techniques; ❐ Material Requirement Planning (MRP); ❐ Enterprise Resource Planning (ERP)

4.1. INVENTORY CONTROL (MEANING AND DEFINITIONS)

Inventory

It is a detailed list of movable goods such as raw materials, materials in process, finished products, general supplies and equipments, etc., and gives the quantity and value of each item.

Inventory Control

It may be defined "as the systematic location, storage and recording of goods in such a way that desired degree of service can be made to the operating shops at minimum ultimate cost".

The Need of Inventory Control. The necessity of inventory control is to maintain a reserve (store) of goods that will ensure manufacturing according to the production plan based on sales requirments and the lowest possible ultimate cost. Losses from improper inventory control include purchase in excess than what needed, the cost of slowed up production resulting from material not being available when wanted. Each time a machine is shut down for lack of materials or each time sale is postponed or cancelled for lack of finished goods a factory loses money.

To promote smooth factory operation and to prevent piling up or idle machine time proper quantity of material must be on hand, when it is wanted. Proper inventory control can reduce such losses to a great extent.

4.2. OBJECTIVES OF INVENTORY CONTROL

Inventory control aims at keeping track of various inventories and this is achieved by the following objectives :

(*a*) Purchasing material at an economic rate, at proper time and in sufficient quantity so as not to run short of them at any instant.

(*b*) Providing a suitable and secure storage location.

(*c*) Providing enough storage space.

(*d*) A definite inventory identification system and adequate and responsible store room staff.

(*e*) Suitable requisition procedure.

(*f*) Up-to-date and accurate record keeping.

(*g*) Periodic inventory verification.

(*h*) Division of inventory under proper heads, exercising the control accordingly and removing obsolete inventory.

In context of inventory management, the firm is faced with the problem of meeting two conflicting needs :

(*i*) To maintain a large size of inventory for efficient and smooth production and sales operations.

(*ii*) To maintain a minimum investment in inventories to maximise profitability.

4.3. QUANTITY STANDARDS

There are five important quantity standards used as tool to control Inventory. These are as follows:

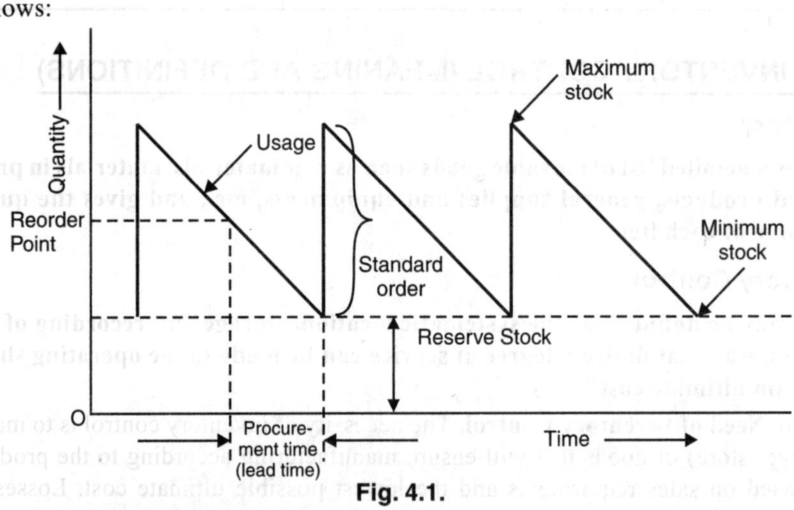

Fig. 4.1.

(*a*) The Maximum Stores.

(*b*) The Minimum Stores.

(*c*) The Standard Order.

(*d*) The Ordering Point.

(*e*) Lead or Procurement Time.

(*a*) **The Maximum Stores.** This term is applied to designate the upper limit of the inventory and represents the largest quantity which in the interest of economy should generally be kept in stores.

(*b*) **The Minimum Stores.** This term is applied to designate the lower limit of the inventory and represents a reserve or margin of safety used in case of emergency. When requirements have been abnormal, it is intended that there must always be atleast this quantity available in stores.

(*c*) **The Standard Order.** It is the quantity to be purchased at any time. Repeat order for a given product are always for this quantity until this is revised.

(*d*) **The Ordering Point.** This represents the quantity required to ensure against exhaustion of the supply during the interval between the placement of an order and receiving delivery.When the balance falls to this level, it is an indication that a new purchase order must be placed.

(*e*) **Lead or Procurement Time.** It is the time which takes the stock to reach from Re-order point to Minimum Stock level. **It may also be defined as the time that elapses between the voicing of a need for an item in production and the time taken to staisfy the need.** If one order is placed today and after 40 days it is fulfilled. Then 40 days period is lead time. Hence it shows that the order should be placed 40 days earlier. **Lead time determines the amount of materials to be kept in reserve.** As the lead time decreases, the reserve stock also decreases and vice versa. Therefore, the lead time analysis is very necessary, and the attempts should be made to reduce this period.

Lead time includes:

(*i*) Time to process the enquiries and to place the order

(*ii*) Time to deliver the order to supplier

(*iii*) Time for the supplier to fulfill the order

(*iv*) Transportation times to reach the purchaser.

In setting maximum, minimum, and order quantities each item should be considered separately in terms of the following factors :

(*a*) Economic size of each purchaser order.

(*b*) Increased lock-up of capital.

(*c*) The time required to receive the goods after requisitioning.

(*d*) The probable depreciation and obsolescence.

(*e*) The rate of demand etc.

4.4. ECONOMIC ORDER QUANTITY (EOQ)

4.4.1. Introduction

The inventory problems in which demand is assumed to be fixed and completely predetermined are usually referred to as the Economic Order Quantity (EOQ). By the order quantity, we mean the quantity produced or procured during one production cycle. When the size of order increases the ordering costs (cost of purchasing, inspection etc.) will decrease whereas the inventory carrying cost (cost of storage, insurance etc.) will increase. Thus in the production process there are two opposite costs; one encourages the increase in the order size and the other discourages. **EOQ is that size of order which minimises total annual (or any other time period as specified by individual firms) costs of carrying inventory and cost of ordering**. The two opposite costs can be shown graphically by plotting them against the order size as shown (Fig. 4.2).

It is evident from above that the minimum total cost occurs at the point where the ordering costs and inventory carrying costs are equal.

Before calculating economic order quantity, it is necessary to become familiar with terms like maximum inventory, minimum inventory, reorder poit etc. Figure 4.3 shows various quantity standards.

Starting from an instant when inventory *OA* is in stores, it gets consumed gradually in quantity from *A* along *AD* at a uniform rate. It is preknown that it takes *L* number of days between initiating order and receiving the required inventory. Therefore, as the quantity reaches point *B*, purchase requisition is initiatd which takes from *B* to *C*, that is time *R*. From *C* to *D* is the inventory procurement time *P*. At the point *D* when only reserve stock is left, the ordered quantity is supposed to reach and again the total quantity shoots to its maximum value, *i.e.,* point *A'* (*A* = *A'*).

Maximum Quantity *OA* is the max. limit to which the inventory can be kept in stores at any time.

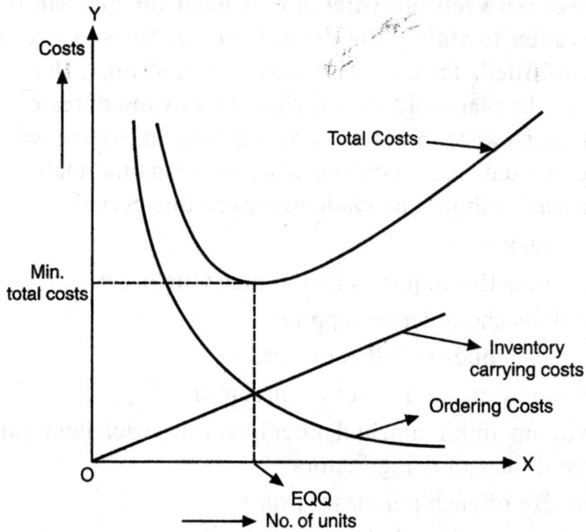

Fig. 4.2. Relationship between cost and quantity

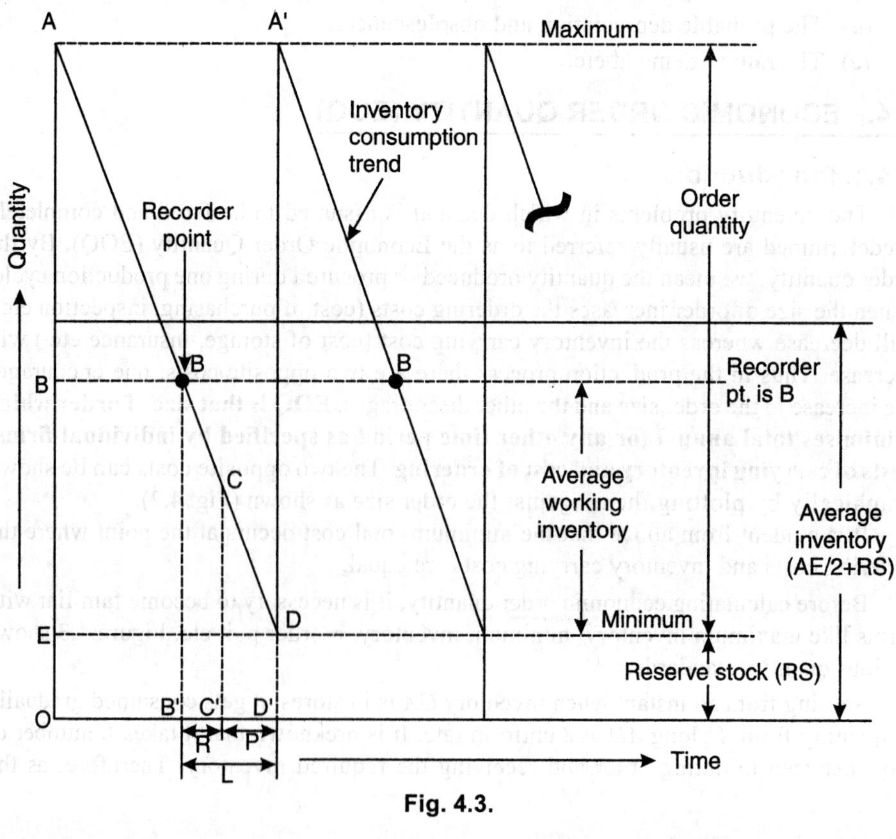

Fig. 4.3.

Minimum Quantity OE is the minimum limit of inventory which must be kept in stores at any time. This is also knowsn as reserve stock (RS) or buffer stock.

Reorder point B indicates that it is a high time to initiate a purchase order and if not done in time, the inventory may exhaust and even reserve stock utilized before the fresh stock arrives. From B' to D' it is the lead time $L = R + P$ and it may be ascertained on the basis of past experience and market condition.

Now let Q = Economic lot size (EOQ)

D = Demand rate; units per year

A = Ordering cost; ₹/order

C = Unit cost; ₹/unit

H = Inventory carrying cost per item per year

TC = Total annual cost of operating the system (₹/year)

$$\text{No. of purchase orders to be furnished} = \frac{D}{Q}$$

$$\text{Total procurement cost} = \frac{D}{Q} \times A$$

$$\text{Average annual inventory} = \frac{Q}{2}$$

$$\text{Total cost} = \text{TC} = \left(\frac{D}{Q} \times A\right) + \left(\frac{Q}{2} \times H\right).$$

To minimise the total cost differentiating with respect to Q we have

$$\frac{d(TC)}{dQ} = -ADQ^{-2} + \frac{1}{2}H = 0 \text{ or } Q° = \sqrt{\frac{2AD}{H}}.$$

$$\text{Differentitating once again } \frac{d^2(TC)}{dQ^2} > 0 \text{ for } Q = \sqrt{\frac{2AD}{H}}, \textit{ i.e.,}$$

$$\text{TC is minimum when } Q = Q^o = \sqrt{\frac{2AD}{H}}.$$

{At EOQ, Ordering or Procurement Cost = Inventory Carrying Cost}

The optimum value of Q has thus been obtained and is given by $Q° = \sqrt{\frac{2AD}{H}}$. Putting the value in total cost above, (Q^{optimal} is represented by Q^o) Minimum cost $= \sqrt{2AHD}$ = Min. total annual inventory cost.

The above formula for EOQ is known as **Wilson's Lot Size formula**. (The most classical of the inventory models was first proposed by Harris in 1915 and further developed by Wilson in the year 1928).

Some interesting insight may be obtained using this classical system.

(*a*) If ordering cost is of high tendency, the optimal policy is to have high EOQ thus raising average inventory level.

(*b*) If the value of H is high, the tendency will be to go for smaller lot sizes.

Assumptions of Basic EOQ Model and their Limitations

Wilson's square root model assumes:

(*a*) Precise knowledge of demand of items and their usage rates are also assumed to be constant.

(*b*) Delivery of units ordered is virtually instantaneous. This assumption is equivalent to saying that cost of not meeting the requirement for all the units ordered is infinite.

(*c*) Price per unit, ordering and carrying costs are all assumed to be constant regardless of order quantity.

(*d*) Stockout costs measured in terms of frustration, delay and embarassment are assumed to be inifinite.

With the above limitations, concept of EOQ is easy to understand and apply in practice. It is also scientifically developed by balancing two opposing costs (ordering cost and inventory carrying cost). The EOQ formula, takes care of most of the above limitations and its vital importance lies in its robustness.

Example 4.1. A cast iron bracket has a demand of 9000 units per year. The cost of one procurement is ₹ 100 and the holding cost per unit is ₹ 2.40 per year. The replacement is instantaneous and no shortage are allowed.

Determine:

(*a*) the economic lot size.
(*b*) the number of orders per year.
(*c*) the time between orders.
(*d*) the company operates for 300 days per year. The procurement time is 10 days and the safety stock is maintained at a level of 100 units. Find the reorder point, the max inventory and the average inventory.
(*e*) minimum total variable yearly cost.
(*f*) the total cost per year if ₹ 3 is the cost of one item.

Solution. (*a*) D = 9000 units/year. A = ₹ 100 per procurement, H = ₹ 2.40/unit/year.

$$Q^\circ = \sqrt{\frac{2AD}{H}} = \sqrt{\frac{2\times100\times9000}{2.40}} = 866 \text{ units per procurement. (rounded off)}$$

(Operating doctrine is to order 866 units in a lot)

(*b*) Number of orders per year $= \frac{D}{Q^\circ} = \frac{9000}{866} = 10.4$ orders/year

(*c*) Time between orders $= \frac{1 \text{ year}}{10.4} = \frac{1}{10.4}\times300 = 29$ days

(*d*) Reorder level = Safety stock + consumption during lead time

$$= 100 + 10 \times \left(\frac{9000}{300}\right) = 400$$

Maximum inventory $= Q^\circ$ + Safety stock

$= 866 + 100 = 966$

Minimum inventory = Safety Stock = 100

Average inventory $= \frac{966+100}{2} = 533$

(*e*) Minimum total variable yearly cost

$$= \sqrt{2ADH} = \text{Min. total annual inventory cost.}$$

$$= \sqrt{2 \times 100 \times 9000 \times 2.40} = ₹\ 2080$$

(*f*) Total cost per year $= 3 \times 9000 +$ Min. total variable yearly cost.

$= 27000 + 2080 = 29{,}080$ (₹) **Ans.**

4.4.2. Types of EQO

The evaluation of the most economic quantity to be purchased involves calculation of the following two costs :

(*a*) Procurement Cost or Buying Cost : This cost includes the expenditure made on:

(*i*) Calling quotations.
(*ii*) Processing quoatations.
(*iii*) Placing purchase orders.
(*iv*) Receiving and inspecting.
(*v*) Verifying and payment of bills.
(*vi*) Other incidental charges, etc.

Set up cost in case of manufacturing :

It includes :

(*i*) Cost of setting up the process or equipment etc.
(*ii*) Cost of scrap which may occur at the beginning of new operation.
(*iii*) Cost of planning production and controlling.
(*iv*) Cost of machine idle time during set up, etc.

(*b*) Inventory Carrying Cost : This consists of expenditure made for :

(*i*) Insurance.
(*ii*) Storage and handling.
(*iii*) Obsolescence and depreciation.
(*iv*) Deterioration.
(*v*) Taxes.
(*vi*) Interest, etc.

This cost varies between 10 to 20% of the product cost.

4.5. SELECTIVE INVENTORY MANAGEMENT

4.5.1. Role of Selective Inventory Control

One of the major operating difficulty in the scientific inventory control is an extremely large variety of items stocked by various organisations. These may vary from 10,000 to 100,000 different types of stocked items and it is neither feasible nor desirable to apply rigorous scientific principles of inventory control in all these items. Such an indiscriminate approach may make cost of inventory control more than its benefits and therefore, may prove to be counter-productive. Therefore, inventory control has to be exercised selectively. Depending upon the value, criticality and usage frequency of an item we may have to decide on an appropriate type of inventory policy. The selective inventory management, thus plays a crucial role so that we can put our limited control efforts more judiciously to the more significant group of items. In selective management we group items in few discrete categories depending upon value ; criticality and usage frequency. Such analyses are popularly known as ABC, VED and FSN Analysis respectively. This type of grouping may well form the starting point in introducing scientific inventory management in an organisation.

4.5.2. ABC (Always Better Control) Analysis

This is based on a very universal Pareto's Law that in any large number we have 'significant few' and 'insignificant many'. For example, only 20% of the items may be accounting for the 80% of the total material cost annually. These are the significant few which require utmost attention.

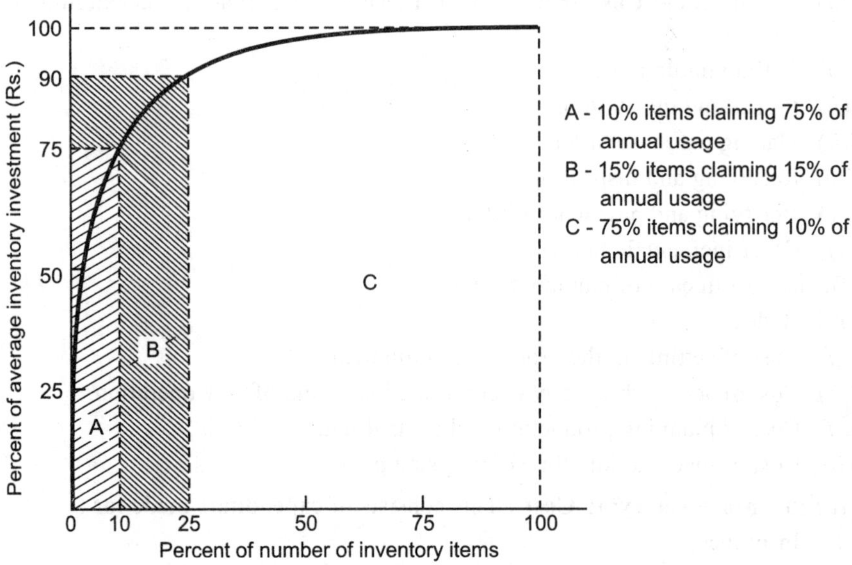

Fig. 4.4. ABC Analysis

Figure 4.4 shows a typical *ABC* analysis showing percentage of number of inventory items and percentage of average inventory investment (annual usage value). Annual usage value is the demand multiplied by unit price, thus giving monetary worth of annual consumption. It can be seen from this figure that 10% items are claiming 75% of the annual usage value and thus constitute the 'significant few'. These are called A-class items. Another 15% items account for another 15% annual usage value and are called B-class items. A vast majority of 75% items account for only 10% expenditure on material consumption and constitute 'insignificant many' and are called C-class items. To prepare an *ABC* type curve, we may follow the following simple procedure :

(*i*) Arrange items in the descending order of the annual usage value.

Annual usage value = Annual demand × Unit price.

(*ii*) Identify cut off points on the curve when there is a perceptible sudden change of slope or alternatively find cut off points at top 10% next 20% or so but do not interpret these too literally — rather as a general indicator.

A very simple empirical way to classify items may be adopted as follows :

$$\text{Average annual usage value } \overline{X} = \frac{\text{Total material cost per year}}{\text{Total number of items}}$$

A-Class items $\geq 6\,\overline{X}$

C-Class items $\leq 0.5\,\overline{X}$

In between we have *B*-class items.

Once the items are grouped into *A, B* and *C* category, we can adopt different degree of seriousness in our inventory control efforts. A class items require almost continuous and rigorous control. Whereas *B*-class items may have relaxed control and *C*-class items may be procured using simple rules of thumb, as usual.

4.6. PURCHASING

4.6.1. Meaning and Definition

Purchasing is the first phase of materials management. Procurement is a function responsible for getting the materials, supplies and equipments of right quality, in the right quantities from the right source, at the right prices and at the right time popularly known as the **five R's** of the art of efficient purchasing.

A company's procurement function becomes particularly important when :

(*i*) Its purchased items account for a high proportion of the unit cost of the product.

(*ii*) When the prices fluctuate widely.

(*iii*) When numerous diverse items are needed and

(*iv*) When the quality of the material appreciably influences the cost of manufacturing.

In the words of Alford and Beatty, "Purchasing is the procuring of materials, supplies, machines, tools and services required for equipment, maintenance and operation of a manufacturing plant."

According to Walters, purchasing function means "the procurement by purchase of the proper materials, machinery, equipment and supplies for stores used in the manufacture of a product adopted to marketing in the proper quality and quantity at the proper time and at the lowest price consistent with quality desired."

Purchasing in an enterprise has now become a specialised function. Any negligence in the purchase function increases the cost resulting into decrease in profits. It was experienced that by giving the purchase responsibility to a specialist, the firm can obtain greater economies in purchasing.

According to Westing, Fine and Zenz, "Purchasing is a managerial activity that goes beyond the simple act of buying. It includes research and development for the proper selection of materials and sources, follow up to ensure proper and timely delivery, inspection to ensure both quantity and quality, development of proper procedures, methods and forms, control traffic, receiving, storekeeping and accounting operations related to purchases".

4.6.2. Classification of Purchases

In a manufacturing firm, the purchases can be classified as under :

(*i*) **Raw Materials.** Raw materials are the materials which are purchased in their natural state which needs further processing and fabrication. Generally raw materials are received from the mine, farm, animal, sea and forest. Iron ore, crude oil, rubber, leather, timber, etc. are certain items of raw materials.

(*ii*) **Components.** Components are finished products which are sub-assembled or assembled to form a finished product. Components are the finished products of the preceding processes and the raw materials for the subsequent operations, *e.g.*, tyres in automobile industries.

(*iii*) **Consumable Stores and Supplies.** These are the items which are required in the manufacturing operations but which normally do not enter into the finished product. Some important supplies are lubricants, abrasives, paper, cleaning compounds, shipping containers, etc.

(*iv*) **Office Supplies.** The office supplies are the indirect materials which are consumed in the office. They include writing paper, carbon paper, stencil paper, duplicating ink and such other materials.

(*v*) **Spares and Tools.** Certain tools and parts are required for repair and maintenance while some tools are required for replacements. They include bolts, nuts, hammer, pumps, wrenches, bearings and different types of tools which are loose and replaceable parts of the machine and equipments.

(*vi*) **Machines and Equipments.** Various types of machines and equipments are required for the manufacturing processes. They include lathes, drill, cutter, furnaces, boilers, trucks, material handling equipments, etc. Their useful life is relatively higher and so, they are treated as capital expenditure items.

4.6.3. Objectives of Purchasing

(*i*) To provide an uninterrupted flow of materials, tools, supplies, etc. to ensure continuous production.

(*ii*) To buy wisely and contribute to the competitiveness of the end product.

(*iii*) To maintain inventories at the optimum level and thus minimize investment in stock.

(*iv*) To bring about better co-ordination with other departments (production, finance, quality control etc.).

(*v*) To develop and maintain good relations with suppliers.

(*vi*) To develop purchase policies and procedures.

(*vii*) To train and develop purchasing personnel working in the organisation.

(*viii*) To select suitable sources of supply and also develop reliable alternative sources of supply.

(*ix*) Collect information about the new materials, tools, equipments, etc. which may reduce the manufacturing costs.

(*x*) To achieve economy and efficiency in the activities of the purchase department by analysing its performance.

4.6.4. Functions of Purchasing

In order to meet the above objectives, the requisite inputs and possible constraints and factors must be identified. The figure given below depicts the purchasing function. Besides the output shown in the figure, the responsibilities of the purchasing department include the following activities:

(*a*) Vendor Rating and Development (*b*) Make or Buy Decision

(*c*) Value Analysis (*d*) Surplus Disposal

(*e*) Control and Audit

(*f*) Maintenance and Development of Procedures, Forms, Records and Reports.

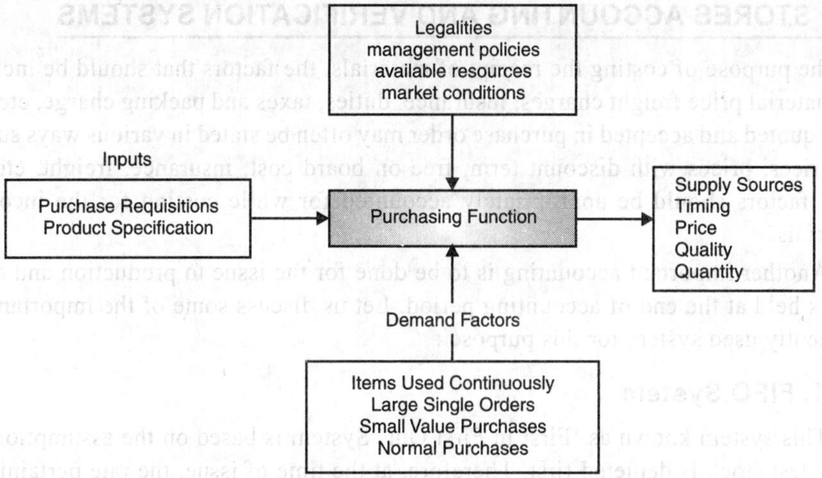

Fig. 4.5. Purchasing Decision Function

4.7. PROCEDURES, FORMS, RECORDS AND REPORTS

Procedure : The following basic steps in which a purchase transactions is carried, from its inception to its conclusion are followed :

(*a*) Receipt and analysis of requisition to assess the need and description of requirement.

(*b*) Selection of possible sources of supply.

(*c*) Determining the time, price, quality and quantity.

(*d*) Placing the order

(*e*) Following up and expediting of the order.

(*f*) Checking the invoice and receiving the order.

(*g*) Processing discrepancies and rejections after inspection.

(*h*) Communicating with accounts section for payment.

(*i*) Closing completed records.

(*j*) Maintenance of records and files.

A schematic purchase procedure involving some of these steps is shown below :

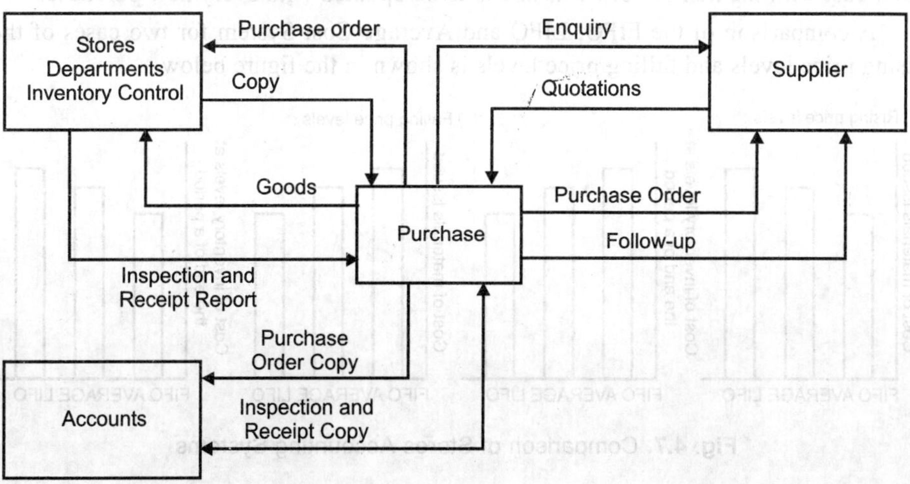

Fig. 4.6. A Schematic Diagram of Purchase Procedure.

4.8. STORES ACCOUNTING AND VERIFICATION SYSTEMS

For the purpose of costing the receipt of materials, the factors that should be included are material price freight charges, insurance, duties, taxes and packing charge, etc. The price quoted and accepted in purchase order may often be stated in various ways such as net prices, prices with discount term, free on board cost, insurance, freight, etc. All these factors should be appropriately accounted for while costing for the incoming materials.

Another important accounting is to be done for the issue to production and of the stocks held at the end of accounting period. Let us discuss some of the important and frequently used system for this purpose :

4.8.1. FIFO System

This system known as 'First in First Out' System is based on the assumption that the oldest stock is depleted first. Therefore, at the time of issue, the rate pertaining to that will be applied. There is no 'profit' or 'loss' in the pricing arrangements. The value of the stocks held on hand is the money that has been paid for that amount of stock at latest price levels. In case of too many changes in price levels, the FIFO System becomes unwieldy. Another limitations of this system is that it fails to provide a satisfactory answer to costing-returns from stores.

4.8.2. LIFO System

This system known as 'Last in First Out' System is based on the assumption that the most recent receipts are issued first. As the latest prices are charged in this system, it leads to lower reported profits in the periods of rising prices and this offers savings in taxes. In case of wide fluctuations in prices, this system tends to immunise unrealised gains or losses in inventory. It has almost the same limitations as that of FIFO System.

4.8.3. Average Cost System

This is based on the assumption that issues to production department are equally made from different shipments in stock, *i.e.* an average cost of shipment in stores is charged. It stabilises the cost figures. The average is to be calculated by dividing the total cost with the number of items and is to be updated with every new purchase.

A comparison of the FIFO, LIFO and Average Cost System for two cases of the rising price levels and failing price levels is shown in the figure below :

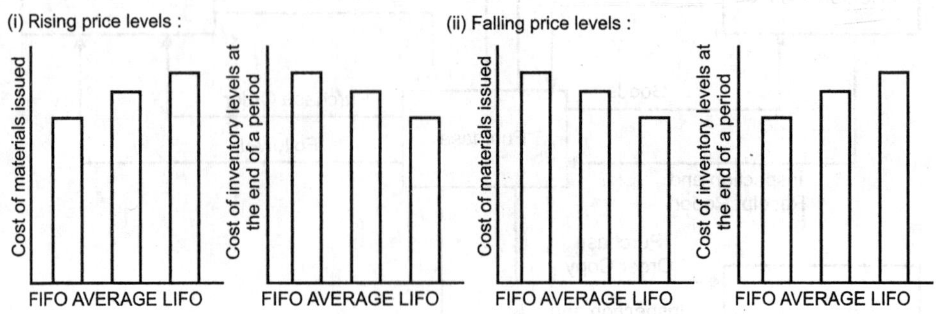

Fig. 4.7. Comparison of Stores Accounting Systems

4.9. MODERN MATERIALS MANAGEMENT TECHNIQUES

4.9.1. Group Technology (GT)

It is the replacement of trsaditional job-shop practice. Here, the operations of the job and their sequences are analysed to form certain families of jobs. Depending on the family, a number of machines are grouped together in each work centre. These work centres are then arranged in a proper sequential order so that these families of jobs flow through the work centres smoothly. The objective here is to minimise the setting time and throughout time.

Group Technology is the realisation that many problems are similar and that by grouping similar problems, a single solution can be found to a set of problems, thus saving time and effort.

This is an improved version of Cellular Layout or Combination Layout (see Cellular layout). In the cellular layout, we observe that at each work centre concerned with one single operation/process say turning, or drilling, etc., some operations are repeated. When more number of products are involved because of variation in sequencing, some products have to backtrack. GT tries to overcome the above limitations of the simple combinations given above. Instead of designing each work centre for a pure, single specialised work, in group technology, each work is a combination of specialised operations or processes. For example, if large number of products have only three operations say drilling, grinding and drilling, (repeated), there is no harm in designing an exclusive work centre catering to these three operations. Thus, each work centre in group technology now becomes a combination of processes instead of being limited to one single process. That particular combination is determined by analysing types of products, their operations and sequencing.

4.9.2. Just-in-Time (JIT)

Think of a situation where we produce the required goods only at the time when they are needed and in the quantity that is needed, and where this holds good for finished products and semi-finished products both. If such a situation materialises, the inventories of the finished goods and work-in-progress would be almost nil. Now if we make our raw materials suppliers agree that they should deliver their goods only at the time and in the quantities we need them to, then we are almost eliminating raw materials inventories as well. We shall then have virtually zero inventories (or near about zero). This is the concept of JIT production system founded by Taiichi Ohno (a Vice President at Toyota) and is first successfully implemented at the Toyota Motor Company's plant in Japan and now being tried at various manufacturing industries all over the world.

JIT is an approach to material management and control designed to produce or deliver parts and materials just when required in production.

Whatever materials and components are needed for a shift are received at the beginning of the shift and converted into finished goods leaving nothing to be carried forward at the end of the shift. It means that the various vendors are working in consonance with the manufacturing organisation as a single team, dedicated to act in unison, wherein production schedule and supply schedules are matched perfectly and operate in an absolute synchronized manner.

4.9.3. Kanban Systems

Kanban system is a kind of production system which operates based on the information contained in cards, called "Kanbans". There are two types of Kanbans namely, Withdrawal

Kanban (WK) and Production Order Kanban (POK). For the purpose of working of the Karbon System, production line is divided into Preceding Process (PP) and Succeeding Process (SP). Withdrawal Kanban contains information on how much material (raw material/semi-finished material) the succeeding process should withdraw. Production Order Kanban contains information on how much quality the preceding processs should produce. Kanban system brings immediately any problem in the system to the notice of the concerned person for corrective action. If a company aims to implement the Kanban system, it should have a well designed maintenance system and motivated workforce.

4.9.4. Traceability

Traceability is the art of having capability to trace something. It is interpreted as the ability to verify the history, location and whereabout of an item in the stores by means of documented recorded identification system. Produce-traceability makes it possible to track down product from its point of origin to a retail location where it is purchased by a customer through the use of a barcode or through traceability software system. Tracing of an item is made easy through various stages of production, manufacturing, processing, handling, transportation, sales and consumption. It simplifies problem of detectation of a defective or contaminated product, ingredient or component. Traceability limits losses and reduces cost.

4.10. MATERIAL REQUIREMENT PLANNING (MRP)

This is generally a computer-based system for drawing up detailed production/ procurement schedules for various parts, sub-assemblies needed to meet a given master schedule of the end item. These are dependable demand items whose demands are linked up with the demand of other product *i.e.* end product. It utilizes the product structure, processing information like production/procurement lead times and inventory status in a bid to produce the best plan. Thus, given a master schedule of end-product, MRP computes the timing of all the sub-assembly, component and raw material production and purchasing activists required over the specified production horizon to meet the master schedule of the final product. Moreover, it does so in such a way as to attempt to minimise work-in-process inventory. Owing to the large amounts of data storage (a typical batch production facility may purchase/produce 20,000 to 1,00,000 components) retrieval and computational requirements, practical MRP systems have to be computerised. **MRP is a system of planning and scheduling the time phased material requirement for production operations.**

Objectives. MRP provides the following objectives :

(*a*) *Inventory reduction.* MRP enables a manager to determine how many of a component are needed and when, in order to meet the master schedule. It avoids the costs of excessive inventory.

(*b*) *Reduction in production and delivery lead times.* MRP identifies materials and component quantities, timings, availability and procurement and production actions required to meet delivery dead lines. MRP helps avoid delays in production. It priorities production activities by putting due dates on customer job orders.

(*c*) *Increased efficiencies.* It provides close coordination among various works centres as products progress through them. MRP focuses on having all components available at appropriately scheduled times. The information provided by MR incourages production efficiencies.

4.10.1. MRP System in Practice

The major advantage of MRP approach is its ability to plan discrete parts production for a very complex production system by scheduling all the activities required to meet the given master-schedule while holding down work-in-process inventory. Many commercially available MRP software packages provide a number of informative report as shown in the figure below :

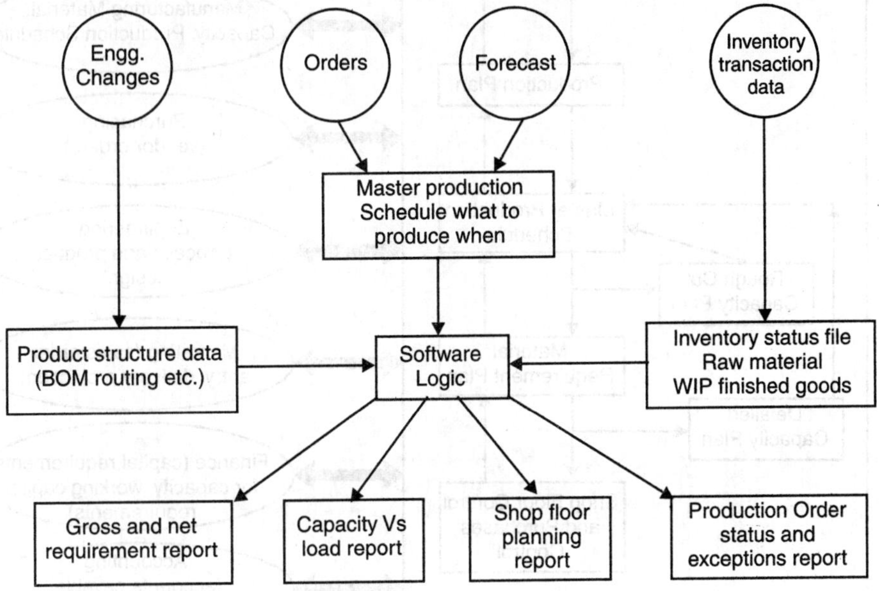

Fig. 4.8. Typical Structure of MRP Based Planning System

The major inputs to the system are:

(*a*) The master production schedule specifying what to produce and when.

(*b*) The product structure data including bill of material, routing files with manufacturing/procurement lead time.

(*c*) The inventory status file for raw material, work-in-process and finished goods.

Among the outputs from the package, the "gross and net requirement report" indicates the timing of order releases required to meet the master schedule.

The "Capacity Vs Load report" determines whether resources required by the work centres are avialable to produce these orders which is useful in planning overtime/sub-contracting.

The "shop floor planning report" is a listing of jobs by due data where the due data indicates when machining at that department should be completed in order to meet the schedule indicated by the gross and net requirements report. The shop-floor supervisor will then release jobs to machines with completion time in hours and due dates for the items.

A typical commercial system provides additional information in the form of "exception report" indicating jobs to be expedited or de-expedited as a consequence of delays in production of components. De-expediting holds down work-in-process inventory.

4.10.2. MRP II (Manufacturing Resource Planning)

It is an integrated information system used to plan and control inventories and capacities in manufacturing companies. MRP II system coordinates sales, purchasing,

manufacturing, finance and engineering by adopting a focal production plan and by using one unified database to plan and update the activities in all the systems.

A schematic diagram of MRP II is given in the figure below :

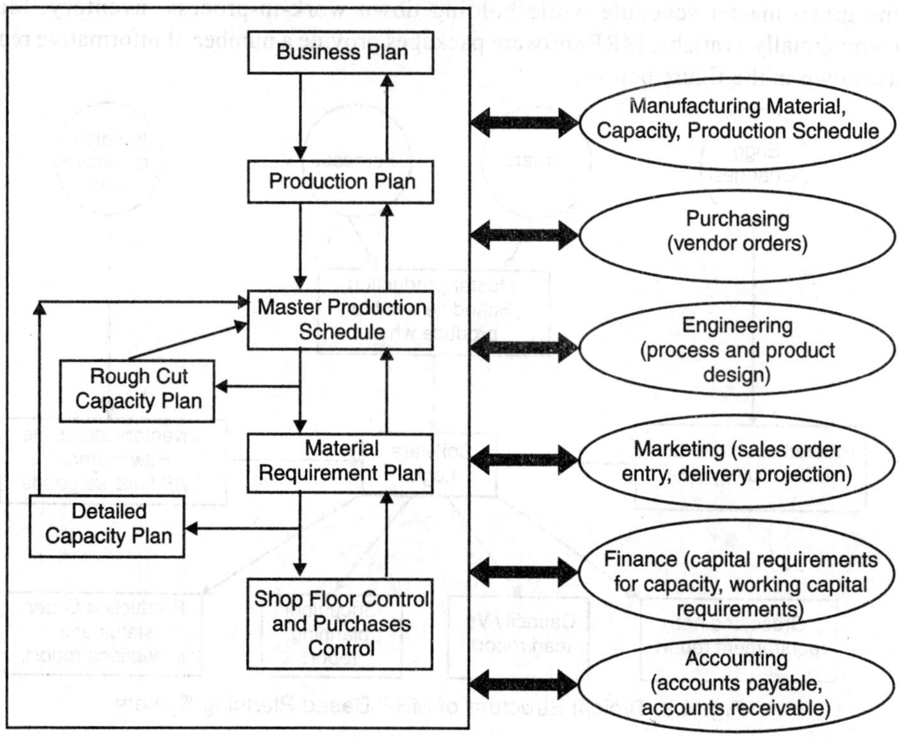

Fig. 4.9. MRP II – An Integrated System for Planning and Control

As shown in Fig. 4.9, the process involves development of production plan from business plan to specify monthly levels of production for each product line over next one to five years. Production plan affect all other functional departments. So, it is developed by all the executives of different functional departments. The production department is expected to produce at the committed levels, the sales department to sell at these levels, and the finance department to ensure adequate financial resources for these levels. Guided by the production plan, the master production schedule specifies the weekly quantities of specific products to be built. At this point a check is made to determine whether the capacity available is adequate to sustain the proposed master schedule. If not, either the capacity or the master schedule must be changed. Once settled, the master schedule is used in the MRP logic to create material requirements and priority schedules for production. Then, an analysis of detailed capacity requirements determines whether capacity is sufficient for producing the specific components at each work center during the scheduled time periods. If not, the master schedule is revised to reflect the limited available capacity. After a realistic, capacity feasible schedule is developed, the emphasis shifts to execution of the plan. From these schedules, work center loading, shop floor control, and vendor follow-up activities can be determined to ensure that the master schedule is met.

4.11. ENTERPRISE RESOURCE PLANNING (ERP)

It is a logical extension of MRP (Material Requirement Planning) and MRP II (Manufacturing resource planning). **ERP is an IT technique to provide the capabilities of collection, generation and storage of vast quantities of data to, an enterprise to automate business functions (such as design, manufacture, distribution, service and recycling of product/services) and enable tracking of information throughout the company whenever required.**

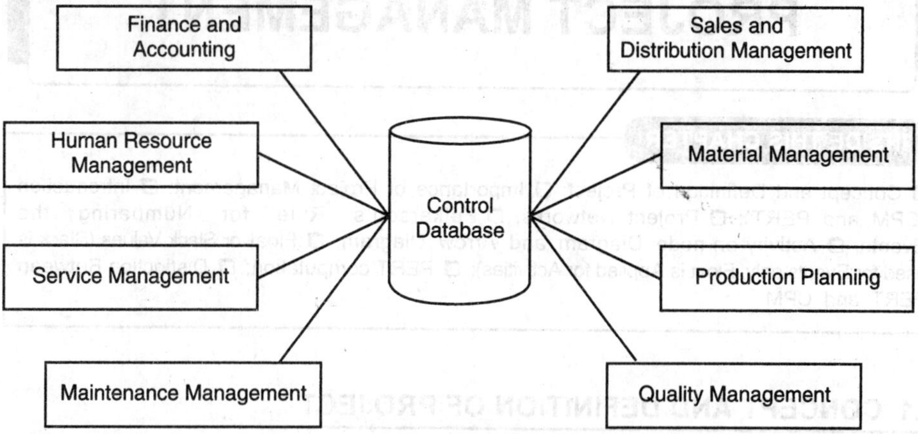

Fig. 4.10. Application modules map of ERP

ERP's core is a single comprehensive database (Fig. 4.10), which collects data from various sources and feeds data into modular applications system supporting all of an enterprise's business functions. It controls all the day-to-day, activities of material management operations (Inventory Control) dealing with stock handling such as goods receipt, goods inspection and goods issue etc.

MODEL QUESTIONS

1. What is inventory control? Explain its importance in an industrial undertaking.
2. Write down the various functions of inventory control.
3. What is an ABC Control Policy? Describe it in details.
4. What do you mean by selective inventory management? Discuss its role critically.
5. What are the criteria commonly employed for the evaluation of purchasing system?
6. "Define purchasing".Write down the various objectives of purchasing.
7. Effective purchasing involves a number of steps. Enumerate them.
8. What is meant by MRP? Explain briefly.
9. What do you understand by manufacturing Resource Planning? Why it is called MRP II? Explain clearly.
10. Explain with a neat sketch the MRP system in practice.
11. Write short notes on:

    (*a*) JIT (*b*) KANBAN

    (*c*) Traceability (*d*) LIFO and FIFO

CHAPTER 5

PROJECT MANAGEMENT

INSIDE THIS CHAPTER

❒ Concept and Definition of Project; ❒ Importance of Project Management; ❒ Introduction (CPM and PERT); ❒ Project Networks; ❒ Fulkerson's Rule for Numbering the Events; ❒ Activity-on-node Diagram and Arrow Diagram; ❒ Float or Slack Values (Slack is used for Events only, Float is Applied for Activities); ❒ PERT computation; ❒ Distinction Between PERT and CPM

5.1 CONCEPT AND DEFINITION OF PROJECT

A project consists of inter-related activities which are to be executed in a certain order before the entire task is completed. **The activities are inter-related in a logical sequence which is known as a precedence relationship.** A particular activity of a project cannot be started until all its immediate preceding activities are completed. Some of the typical projects are :

(*a*) Construction of a house

(*b*) Commissioning of a factory

(*c*) Construction of a ship

(*d*) Fabrication of a steam boiler

(*e*) New product launching in the market

(*f*) Research to develop a new technology, etc.

Definition of Project

According to Project Management Institute, U.S.A, **"A Project is a one shot, time limited, goal directed, major undertaking, requiring the commitment of varied skills and resources."**

Characteristics of Project

(*i*) Project is one time activity which never be repeated exactly in the same manner.

(*ii*) A project has a definite start and finish i.e., a project is executed in a definite time bound schedule.

(*iii*) A project uses a cross-sectional relationship because it needs diversified skills and talents from different professions.

(*iv*) A project has definite goals and end results that can be defined in terms of cost, schedule and performance requirements.

(*v*) Project demands the investment (current outlay) and the benefits are spread for number of future periods.

(*vi*) Once the project goals are achieved, the project team will be either disbanded or reconstituted for another new project.

(*vii*) Project passes through several distinct activities which constitute a project life cycle.

5.2. IMPORTANCE OF PROJECT MANAGEMENT

Project Management is concerned with defining project activities (planning, organizing, scheduling and controlling) maintaining their integrity and ensuring that they are performed as desired within the allocated time and cost budget, using organizational and non-organizational resources.

It is generally applied for constructing public utilities, large industrial projects and organizing mega events. Project Management is considered to be an important field in production scheduling mainly because many of the industrial activities can also be viewed as project management problems. Many modern projects involve great technical complexities and requires much diversity of skills and to cope up with complex kind of activities, new form of project organizations and new practices of management have evolved. Project management is one of the most challenging and creative tasks.

Projects are important for individuals organizations and economies :

(*i*) For individuals project management can provide one of the most challenging career paths.

(*ii*) For organizations, there is the opportunity to derive competitive advantage from their projects.

(*iii*) For nations, the performance of projects will have an impact on their economies.

Let us go back to the older days to study the importance of project management and turn the pages of history. During the days of World War I, Henry Gantt developed an effective tool for the project work – Gantt chart – which is still used widely with advantage. During 1950s, formal tools and techniques were developed to manage large complex projects which were risky and uncertain. In 1957. Du Pont Corporation developed CPM (Critical Path Method) for scheduling company's production facilities (Project work) and in 1958 Rand Corporation developed PERT (Project Evolution and Review Technique) for planning missile development. By the end of the 5th decade of twentieth century, installations of computerized data processing systems provided immense amount of information to manage large scale projects. It was in 1970s that the concept of Performance Measurement System came into existence which led to more reliable forecasting of final project costs and completion dates. Today a large variety of quality management software programmes are available to expand the planning and control capabilities of managers even for small projects. The ready availability of technology in modern times is virtually the outcome of successful pursuance of projects over a period of so many years. In today's world of shortened project life cycles, project management skills are more important than ever. The future is bright and more result oriented with the new crops of project managers to make maximum mileage with the limited resources.

5.3. INTRODUCTION (CPM AND PERT)

In the present era of complexity of problems of management like availability of skilled manpowers, timely input of raw materials, near absence of improved technology and on the top of it scarce money power dogs the completion of a project in time to the satisfaction of the authorities. The sum total of the odds has added to the pressure on the brains of decision-makers. Management are, thus, always on the look out for some methods and techniques which may help in planning and scheduling in controlling the project. The aim of the planning is to develop a sequence of activities of the project so that the project completion time and cost are properly balanced and wastages are minimized. To meet the objectives of systematic planning, a number of techniques have been evolved applying network strategy. Having got recognition in the year 1955-56, when dramatic reductions in time (about 70%) were experienced in connection with the overhaul of generating plant by Central Electricity Generating Board in Great Britain, the two techniques are **PERT (Project Evaluation and Review Technique) and CPM (Critical Path Method).**

The technology travelled to India crossing the seven seas about five years, hence. Since then a number of beneficiaries like HMT, NTPC, ONGC, Tata Group and a host of other industrial houses harnessed power by solving complex problems of management effectively. **The techniques of PERT and CPM are essentially network scheduling techniques used for planning and controlling projects in the field of construction, maintenance, fabrication, purchasing and quality control, etc. They employ the technique of minimizing trouble spots such as production bottlenecks, delays and interruptions by determining the critical factors and coordinating various parts of overall job. We shall deal with the applications of CPM first.**

5.4. PROJECT NETWORKS

1. Network and Basic Components: A network is a graphic representation of a project's operations and is composed of activities and events that must be completed to reach the end objective of a project showing the planning sequence of their accomplishments, their dependence and inter-relationship.

2. Event : An Even (or node) is a specific physical or intellectual accomplishment in a programme or project plan. It is a particular instant of time and does not consume time or resource.

3. Activity : It is a task or item of work to be done, that consumes time, effort, money or other resources. It lies between two events called the preceding and succeeding ones. (See Fig. 5.1). In the network construction, the length of the activity may not be proportional to the time duration of the activity. For example, an activity with a time duration of 5 days (say) may be shown equal to or less than or more than the activity with time duration of 1 day (lengthwise). The length and direction of the activity are only symbolic of the time period showing the direction of progress of the activity and the length is so chosen to complete the construction of the network conveniently irrespective of the actual time duration of the activity.

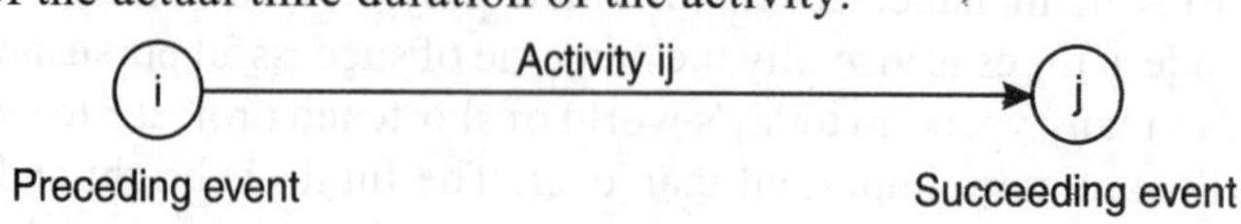

Fig. 5.1.

4. Dummy Activity : An activity which only determines the dependency of one activity over the other, but does not consume any time is called a dummy activity. It is represented by dotted line arrow. Dummies generally serve the following purposes:

(*i*) To maintain uniqueness in the numbering system, as every activity may have a distinct set of events by which the activity can be identified (a dummy activity has a zero time).

(*ii*) To maintain the proper logic in the network.

Suppose activities *B* and *C* have activity *A* as their immediate predecessor. It will be shown as shown below :

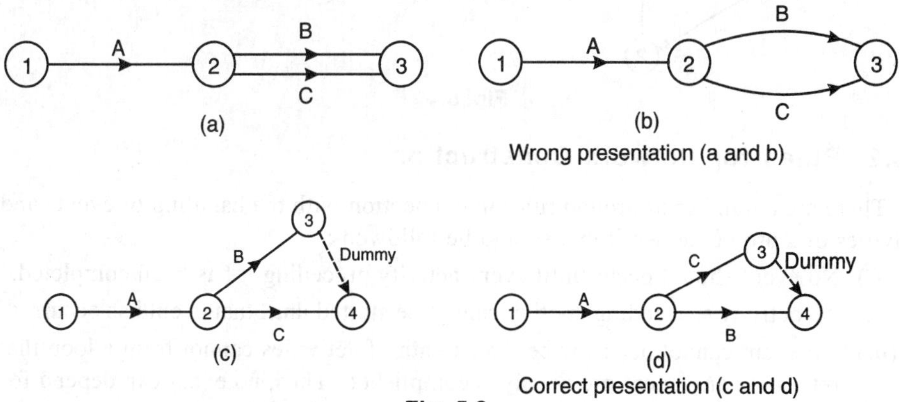

Fig. 5.2.

5.4.1. Common Errors in Network Drawing

There are three types of errors which are most common in network drawing, viz.,

(*i*) **Formation of Loop:** If an activity were represented as the Fig. 5.3 (*a*) below with activity B reversed in direction. Cycling in a network can result through a simple error or when, while developing the activity plan, one tries to show the repetition of an activity before beginning the next activity. A closed loop will produce an endless cycle in computer programmes without a built-in routine for detection or indentification of the cycle. Thus one property of a correctly constructed network diagram is that it is "non-cyclic".

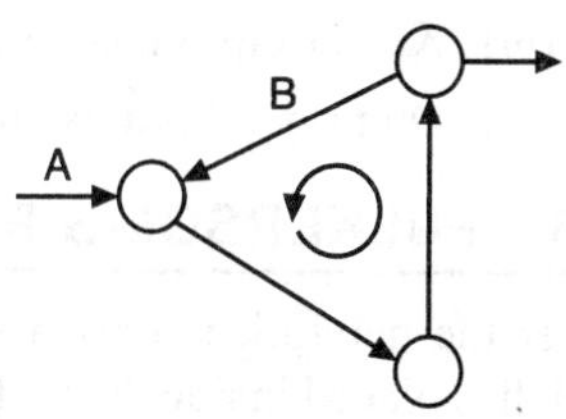

Fig. 5.3. (a) Closed loop

(*ii*) **Dangling:** No activity should end without being joined to the end event. If it is not so, a dummy activity is introduced in order to maintain the continuity of the system. Such end-events other than the end of the project as a whole are called dangling events. In the Fig. 5.3 (*b*) activity *D* leads to dangling. A dummy activity is therefore introduced to avoid dangling. Here dummy D_1 is used to avoid dangling.

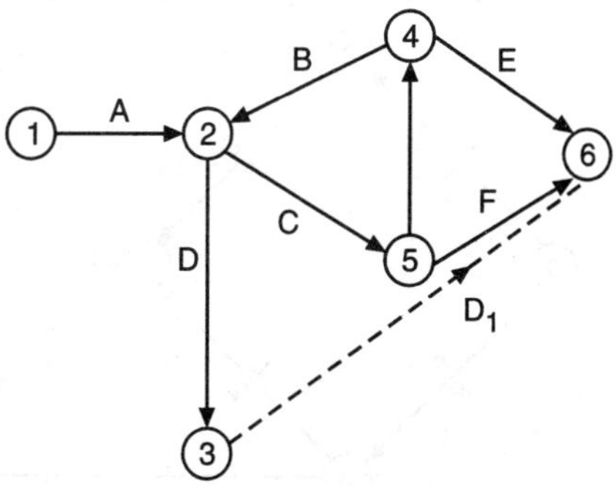

Fig. 5.3. (b) Dangling event

(*iii*) **Redundancy:** If a dummy activity is the only activity emanating from an event, it can be eliminated. For example, in the network shown in Fig. 5.4 the dummy activity is redundant and can be eliminated and the network redrawn.

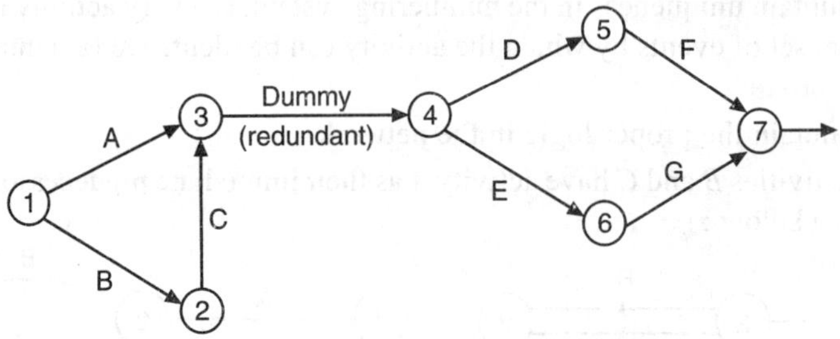

Fig. 5.4

5.4.2. Rules for Network Construction

There are a number of ground rules in connection with the handling of events and activities of a project network that should be followed :

(*i*) No event should occur until every activity preceding it has been completed.

(*ii*) An activity succeeding another cannot be started until that event has occurred.

(*iii*) An event cannot occur twice, *i.e.*, a path of activities cannot form a loop that returns to any event previously accomplished. Thus, no event can depend for its completion upon the completion of a succeeding event.

(*iv*) Each activity must start from and terminate in an event.

(*v*) Time flows from left to right.

(*vi*) An activity must be completed in order to reach the end-event.

(*vii*) Dummy activities should only be introduced, if absolutely necessary.

5.5. FULKERSON'S RULE FOR NUMBERING THE EVENTS

After the network is drawn in a logical sequence, every event is assigned a number which is placed inside the node circle. The number sequence should be such as to reflect the flow of the network. The rule deviced by D.R. Fulkerson is used for the purpose of numbering. The steps are the following :

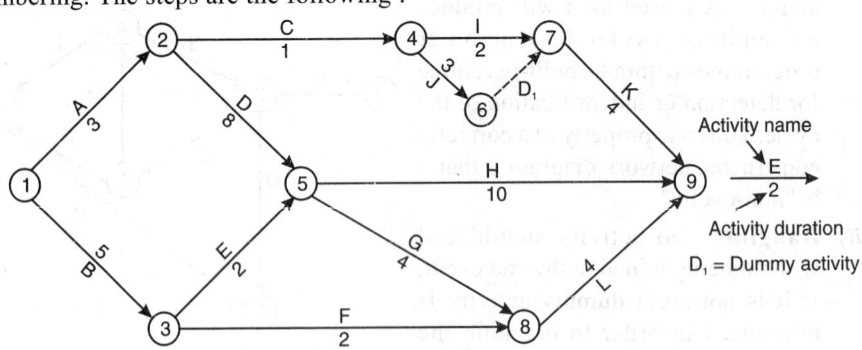

Fig. 5.5.

CPM and PERT

(*i*) Event number should be unique.

(*ii*) Event numbering should be carried out in a sequential basis from left to right.

(*iii*) The initial event which has all outgoing arrows with no incoming arrow is numbered 0 or 1.

(*iv*) The head of an arrow should always bear a number higher than the one assigned at the tail of the arrow. For example – 1-2, 1-3, 2-4, 2-5, 3-7, 3-8, etc. and not 8-7 or 9-3 and the like.

(*v*) Gaps should be left in the sequence of event numbering to accommodate subsequent inclusion of activities, if necessary.

Given below is a example of a project network drawn as per laid down rules. (See Fig. 5.5)

[Note that all the activities are in progressively increasing order, like 3-5, 5-8, 8-9, 4-7, etc.]

5.6 ACTIVITY-ON-NODE DIAGRAM AND ARROW DIAGRAM

Also called Precedence Diagram, this technique was developed by John W. Fondahl in U.S.A and Bernard Roy in France almost at the same time. In AON diagram, the activities are denoted by nodes while the precedence relationships are represented by the arrows. Arrow diagram is also shown in Fig. 5.6. The following Fig. 5.6 shows an AON diagram (*b*) and an arrow diagram (*a*)

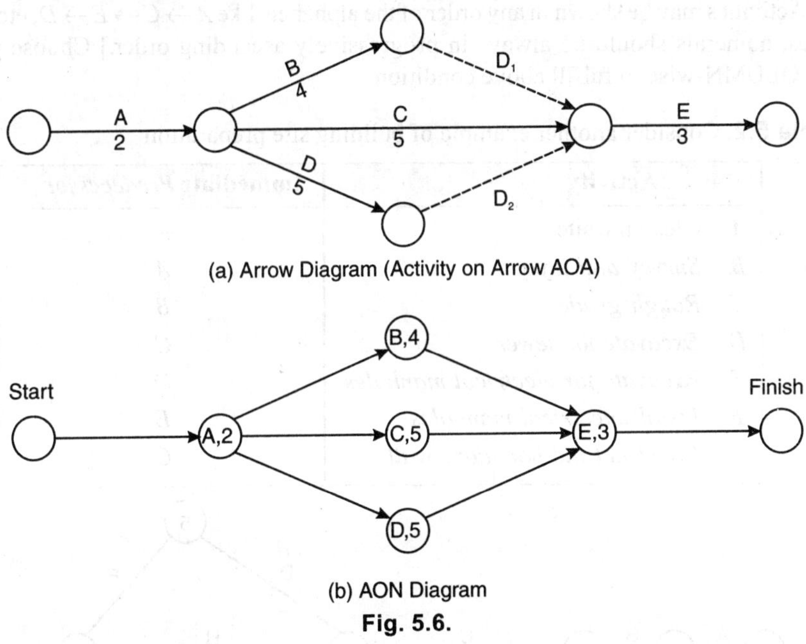

(a) Arrow Diagram (Activity on Arrow AOA)

(b) AON Diagram

Fig. 5.6.

[Arrow Diagram and AON are the two approaches that can be used to determine the project critical path and each one is equally suitable.]

The examples that follow will clearly illustrate the application. (Readers are advised to use pencil and eraser for the construction of network, because they may get the correct construction, initially, in two or three trials or even more. But remember it takes a sharp mind to make a network, not a sharper pencil.)

Example 5.1. (Example of Network Construction). List of activities for a project – Arranging a tea party.

A — *Procure material*
B — *Prepare snacks*
C — *Arrange for the cutlery*

D — *Prepare tea*
E — *Set table with snacks and cutlery*
F — *Serve tea*
G — *Clean cutlery.*

List of Predecessor for Example 1.

Activity	Immediate predecessor	Network (Fig. 5.7)
A	—	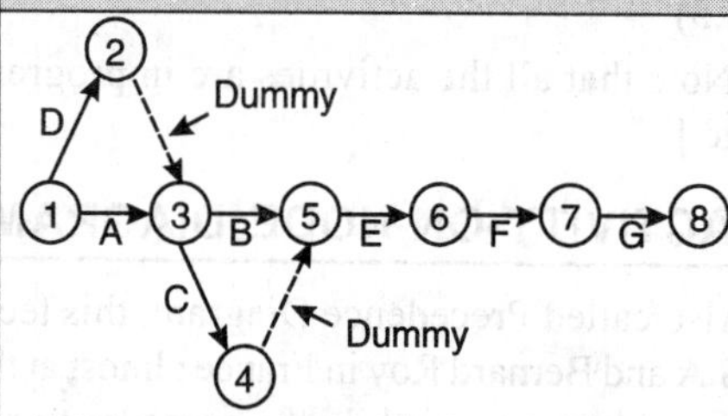
B	A, D	
C	A, D	
D	—	
E	B, C	
F	E	
G	F	

Fig. 5.7.

[**Note :** Activities may be shown in any order of the alphabets like $A \rightarrow C \rightarrow E \rightarrow D$, etc. but on nodes, numerals should be always in progressively ascending order.] Choose your nodes COLUMN-wise to fulfill above condition.

Example 5.2. Consider another example of building site preparation

Activity	Immediate *Predecessor*
A. Clean the site	—
B. *Survey and layout*	*A*
C. *Rough grade*	*B*
D. *Excavate for sewer*	*C*
E. *Excavate for electrical manholes*	*D*
F. *Install electrical manholes*	*E*
G *Construct the boundary wall*	*C*

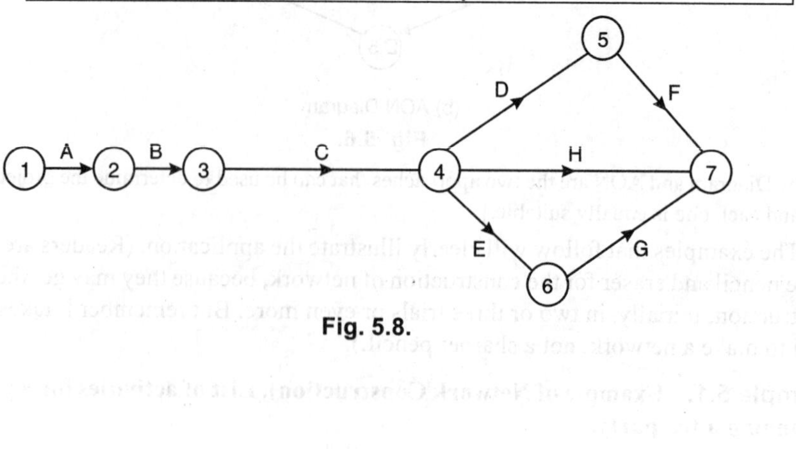

Fig. 5.8.

Example 5.3. Consider a project with the following precedence requirement

Activity/Time duration	*Immediate Predecessors*
$A = 3$	—
$B = 5$	—
$C = 7$	—
$D = 4$	C
$E = 7$	A, B
$F = 3$	E, D
$G = 1$	D
$H = 2$	F, G

Nodes should proceed in ascending order, like 1-2, 1-4, 6-7, ... etc. It should not be 7-5 or 4-3 etc. Network construction is shown step wise in Fig. 5.9.

Critical Path : A path in a network is a sequence of activities which have to be completed sequentially due to precedence requirements. Thus, *A, E, F, H* is a path and so are (C, D, D_2, F, H), (B, D_1, E, F, H), (C, D, G, H), in Fig. 5.9. Time duration of the path is the sum of activity times of the activities forming the path. **The path or paths with the largest time duration in the network is (are) called critical path(s). All the activities on the critical paths are called critical activities.** The earliest time by which a project can be completed is the time durations of the critical path. If an activity on the critical path is delayed the whole project will get delayed. Thus, for timely completion of the project, it is essential that all critical activities be started at the earliest start time and be completed in the planned duration.

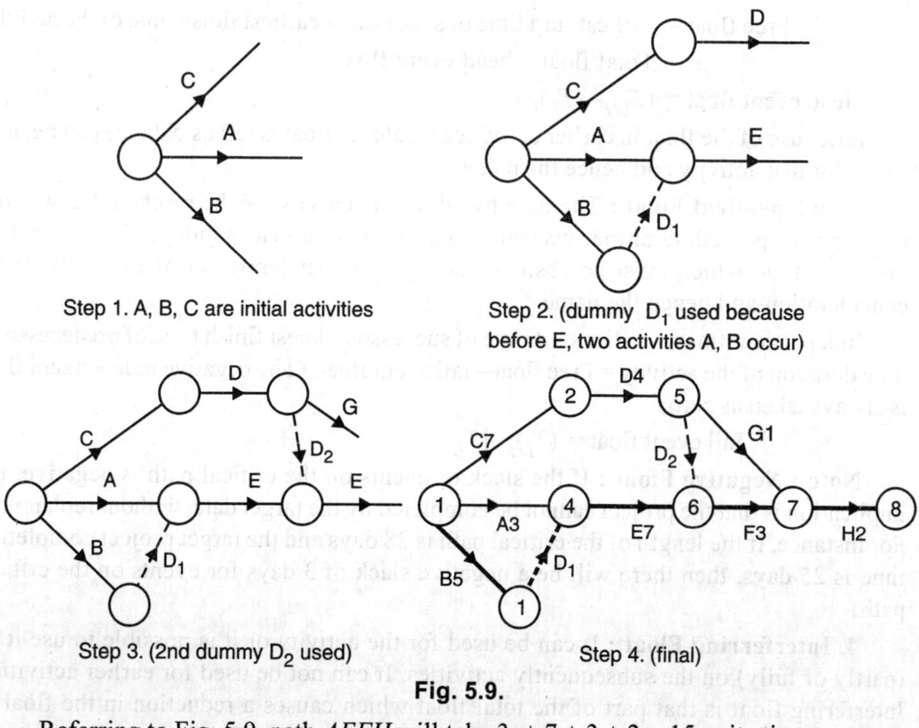

Fig. 5.9.

Referring to Fig. 5.9 path *AEFH* will take 3 + 7 + 3 + 2 = 15 units time.

path – $CDD_2\,FH$ will take 7 + 4 + 3 + 2 = 16 units.

path – BD_1 EFH will take $5+7+3+2=17$ units (longest)

path – $CDGH$ will take $7+4+1+2=14$

So, path BD_1 EFH or simply $BEFH$ having the longest duration is the critical path shown by thick dark lines. The second longest path is called the subcritical path and so, *CDFH* is the subcritical path having next longest duration of 16 units. Other paths are called non-critical paths

5.7. FLOAT OR SLACK VALUES (SLACK IS USED FOR EVENTS ONLY, FLOAT IS APPLIED FOR ACTIVITES)

Float or slack refers to the amount of time by which a particular activity or event can be delayed without affecting the time schedule of the network. There are in general four types of floats :

(1) Total float, (2) Free float, (3) Independent float, (4) Interferring float.

1. Total Float : It is the amount of time by which the completion of an activity could be delayed beyond the earliest expected completion time without affecting the overall project duration time. Difference between latest finish time and earliest finish time of the activity. $lf - ef =$ total float. This is also equal to latest start time – earliest start time of the activity.

Thus, total float $= lf - ef = ls - es$.

2. Free Float : It is the amount of time by which the completion of an activity can be delayed beyond the earliest finish time without affecting the earliest start of a succeeding activity. This is based on the assumption that all activities start as early as possible.

Free float = earliest start time of successor – earliest finish time of the activity

= Total float – head event float.

Head event float = $(T_L)_j - (T_E)_j$

After use of the float in earlier activities whatever float is left as balance can be used *freely* for that activity and hence the name.

3. Independent Float : The time by which an activity can be rescheduled without affecting the preceding or the succeeding activities is known as independent float. It is the spare time which must be absorbed *solely* (Independently) on the activity under consideration and hence the name.

Independent float = earlier start time of successor – latest finish time of predecessor – time duration of the activity = Free float – tail event float. (The negative independent float is always taken as zero)

Tail event float = $(T_L)_i - (T_E)_i$

Note : Negative Float : If the slack of events on the critical path is negative, the implication is that the project cannot be completed by the target date, without replanning. For instance, if the length of the critical path is 28 days and the target project completion time is 25 days, then there will be a negative slack of 3 days for events on the critical path.

4. Interferring Float : It can be used for the activity or it is possible to use it up (partly or fully) on the subsequently activities. It can not be used for earlier activities. Interfering float is that part of the total float which causes a reduction in the float of successor activities. It is the difference between the latest finish time of activity in question and the earliest starting time of the following activity or zero whichever is larger. It represents the portion of the float of an activity which cannot be consumed without adversely affecting (or interfering) the float of the subsequent activities and hence the name. (Interference float = Total float – free float).

Note : Independent float ≤ Free float ≤ total float (See Fig. 5.10 for illustration).

Refer to Fig. 5.15. We take into consideration activity B with duration of 8.

(*i*) We have T_E(tail) + total float + duration = T_L (head)

∴ 4 + total float + 8 = 14

∴ total float = 14 – 4 – 8 = 2

(*ii*) T_E(tail) + free float + duration = T_E (head)

∴ 4 + free float +8 = 12 or free float

= 12 – 4 – 8

= 0 (= total float – head event float)

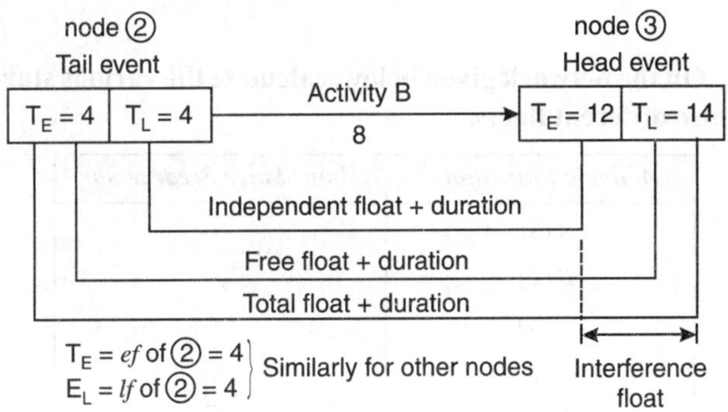

Fig. 5.10

(*iii*) T_L (tail) + Independent float + duration = T_E(head)

or 4 + Independent float + 8 = 12 or

independent float = 12 – 4 – 8

= 0 (free float – tail event float).

(*iv*) Interference float $= T_L - T_E$

= 14 – 12 = 2 (also equal to total float – free float)

Note : Refer Fig. 5.15. We see that the total float of B is 2 days and that of D is 2 days, but it does not mean the total float of path 2-3-5 = 2 + 2 = 4 days. It means that the total float of B = 2 days or total float of D = 2 days and the total float of path 2-3-5 is also 2 days. If B utilizes 1.5 days float alone then the total float of D = 2 – 1.5 = 0.5 days only. The concept of free float is that if B consumes 0.75 days float, then D in the path 2-3-5 is *free* to consume remaining 1.25 days float itself. If we earmark 1 day float each to B and D separately to be *solely* used by them, then this will be *independent float* for each activity. In case B taken full quota of 2 days float to itself, then it adversely affects or *interferes* with the float of the subsequent activity D (which will leave no freedom to D to relax with any float now).

Summary

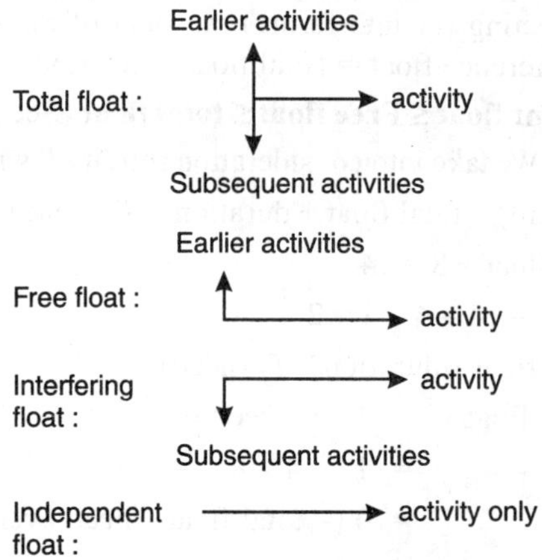

Example 5.4. On the network given below, calculate the various start and times of the activities and also different floats.

Actively (duration)	*Impedance Predecessor*
A (3)	—
B (2)	*A*
C (3)	*A*
D (3)	*C*
E (2)	*C*
F (7)	*B, C*
G (5)	*B, C*
H (6)	*D, F*

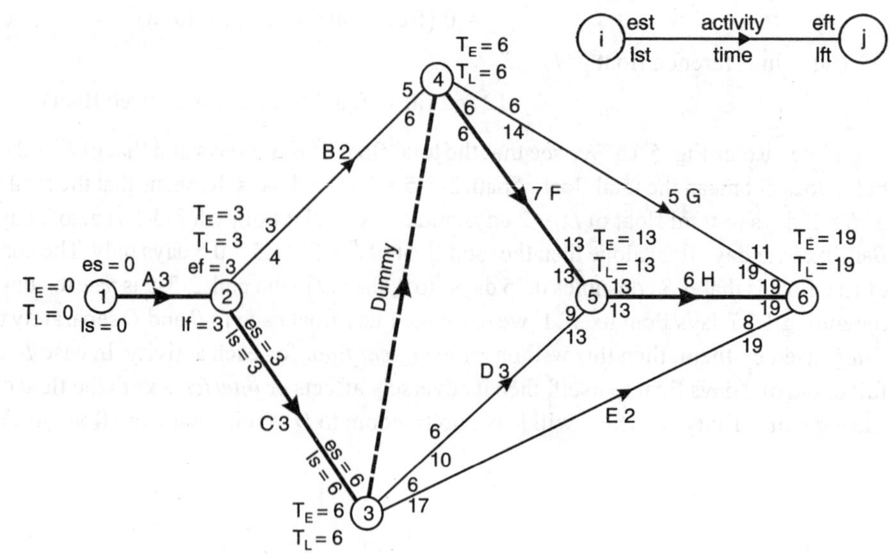

es = early start, ef = early finish, ls = late start, lf = late finish,
T_E = time earliest for the node event, T_L = time latest for the node event

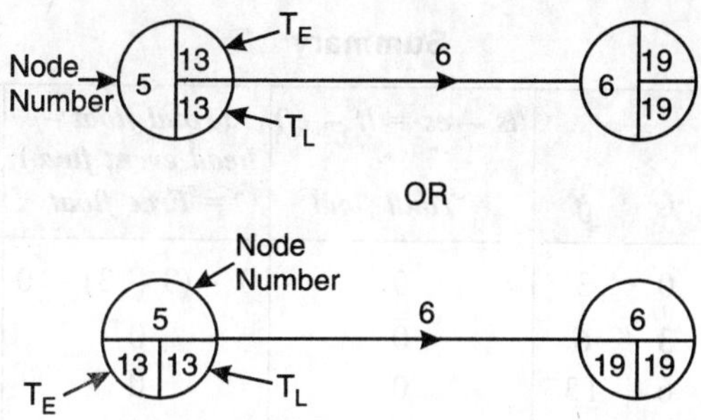

Fig. 5.11.

Paths are $1-2-3-4-5-6 = 3+3+7+6 = 19$ (Critical path)

$1-2-4-6 = 3+2+5 = 10$

$1-2-3-4-6 = 3+3+5 = 11$

$1-2-3-6 = 3+3+2 = 8$

$1-3-3-5-6 = 3+3+6 = 15$ (Sub-critical path)

For computation of early start and early finish time, we start from left to right **(Forward Pass Method)**. For late start and late finish time we start from the end, i.e., from right to the left. **(Backward Pass Method)**

Activity	*Early start*	*Early finish*
(1 – 2)	0	0 + 3 = 3
(2 – 3)	3	3 + 3 = 6
(4 – 5)	6	6 + 7 = 13
(5 – 6)	13	13 + 6 = 19
So, the critical path is covered first.		
(2 – 4)	3	3 + 2 = 5
(4 – 6)	max. of ef_{2-3} and ef_{2-4} i.e., max of 6 and 5 (see note below) that is 6	6 + 5 = 11
(3 – 6)	6	6 + 2 = 8
(3 – 5)	6	6 + 3 = 9

Activity	*Late finish*	*Late Start*
(5 – 6)	19	19 – 6 = 13
(4 – 5)	13	13 – 7 = 6
(2 – 3)	6	6 – 3 = 3
(1 – 2)	3	3 – 3 = 3
Critical path is covered first.		
(4 – 6)	19	19 – 5 = 14
(2 – 4)	min. of ls_{4-5} and ls_{4-6} i.e., min of 6 and 14 (see note below) that is 6	6 – 2 = 4
(3 – 6)	19	19 – 2 = 17
(3 – 5)	13	13 – 3 = 10

Summary

Activity	es	ef	ls	lf	(ls – es = lf – ef) = Total float	(Total float – head event float) = Free float	(Free float – tail event float) = Independent float
1—2	0	3	0	3	0	0 – (3 – 3) = 0	0 – (0 – 0) = 0
2—3	3	6	3	6	0	0	0
4—5	6	13	6	13	0	0	0
5—6	13	19	13	19	0	0	0
2—4	3	5	4	6	1	1	1
4—6	6	11	14	19	8	8	8
3—5	6	9	10	13	4	4	4
3—6	6	8	17	19	11	11	11

Note : Please note the early start time of (4 – 6) as to how it is calculated out. There are 2 activities before (4 – 6), activity (2 – 4) and (2 – 3) and (4 – 6) will only start after each of the activities (2 – 4) and (2 – 3) will finish and that's why we take that time of the activity into reckoning which has finished later and for which time duration is more. We choose the maximum of ef_{2-3} and ef_{2-4}, i.e., max. of 6 and 5 = 6.

Similarly while calculating late finish time of (2 – 4) we observe that 2 activities (4 – 5) and (4 – 6) are after it. Activity (4 – 5) will start latest by day 6 (6th day) and (4 – 6) by day (14th day) and so (2 – 4) must get finished by the minimum time of 6 and 14, and that's why we choose lf_{2-4} as 6 (minimum of 6 and 14). In general, student should bear in mind the following key points:

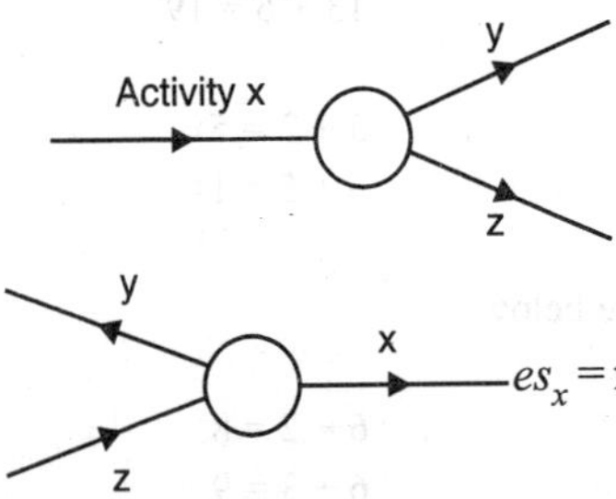

lf_8 = min of (ls of y, ls of z) lf_x = min of (ls of y, ls of z)

es_x = max of (ef of y, ef of z) es_x = max of (ef of y, ef of z)

Fig. 5.12.

Example 5.5. The table below lists all the activities which together constitute a small engineering project. The table also displays the necessary immediate predecessors for each activity.

Activity	–	*A*	*B*	*C*	*D*	*E*	*F*	*G*	*H*	*I*	*J*
Immediate Predecessor	–	–	*A*	*A*	*A*	*B*	*C, D*	*D*	*B*	*EFG*	*G*
Activity Duration (days)	–	2	3	4	5	6	3	4	7	2	3

(*a*) *Calculate earliest start time, finish time, latest start time and finish time of each activity.*

(*b*) *The critical path.*

(*c*) *Total float of each activity.*

(*d*) *Latest start day for activity B.*

(*e*) *The effect on the project duration, if activity I were to take 3 days.*

(*f*) *The effect on the project duration, if activity F were to take 6 days.*

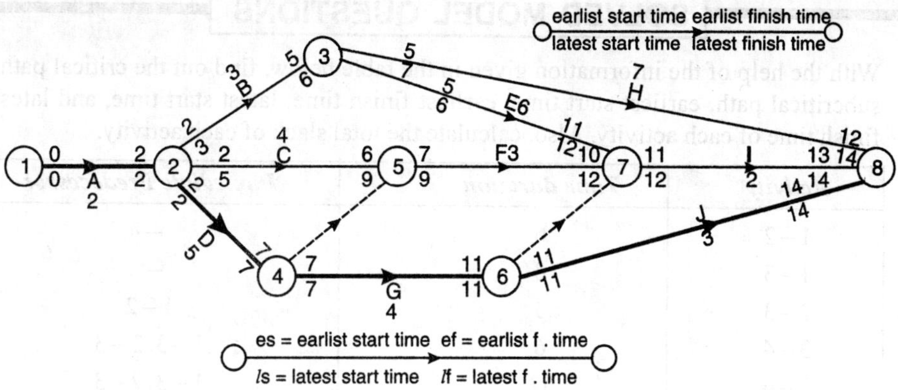

Fig. 5.13.

Path *ADGJ* = 2 + 5 + 4 + 3 = 14 days (Critical path)

Path *ABH* = 2 + 3 + 7 = 12 days

Path *ABEI* = 2 + 3 + 6 + 2 = 13 days (Sub-critical)

Path *ACFI* = 2 + 4 + 3 + 2 = 11 days

Path *ADFI* = 2 + 5 + 3 + 2 = 12 days

Path *ADGI* = 2 + 5 + 4 + 2 = 13 days (Sub-critical)

(*a*)

Part (*c*) of question

Activity	*es*	*ef*	*ls*	*lf*	*Slack* *ls – es = lf – ef*
* A	0	2	0	2	0 – 0 = 2 – 2 = 0
B	2	5	3	6	3 – 2 = 6 – 5 = 1
C	2	6	5	9	5 – 2 = 9 – 6 = 3
* D	2	7	2	7	2 – 2 = 7 – 7 = 0
E	5	11	6	12	6 – 5 = 12 – 11 = 1
F	7	10	9	12	9 – 7 = 12 – 10 = 2
* G	7	11	7	11	7 – 7 = 11 – 11 = 0
H	5	12	7	14	7 – 5 = 14 – 12 = 2
I	11	13	12	14	12 – 11 = 14 – 13 = 1
* J	11	14	11	14	11 – 11 = 14 – 14 = 0

* Activities on critical path.

(*d*) Latest start time for activity *B* = 3 days

(*e*) If activity *I* were to take 3 days instead of 2, then path *ABEI* = 14 days will become critical along with path *ADGJ*.

(*f*) Activity *F* were to take 6 days instead of 3, then path *ADFI* will take 12 + 3 = 15 days, i.e, it will become now critical and the project will get delayed by 1 day.

SOLVED MODEL QUESTIONS

1. With the help of the information given in the table below, find out the critical path, subcritical path, earliest start time, earliest finish time, latest start time, and latest finish time of each activity. Also, calculate the total slack of each activity.

Activity	*Time duration*	*Immediate Predecessor*
1 – 2	6	—
1 – 3	8	—
2 – 3	3	1 – 2
3 – 4	6	1 – 3, 2 – 3
3 – 6	12	1 – 3, 2 – 3
3 – 5	10	1 – 3, 2 – 3
2 – 4	8	1 – 2
4 – 6	8	2 – 4, 3 – 4
5 – 6	6	3 – 5
5 – 7	12	3 – 5
6 – 7	8	4 – 6, 3 – 6, 5 – 6
7 – 8	7	6 – 7, 5 – 7

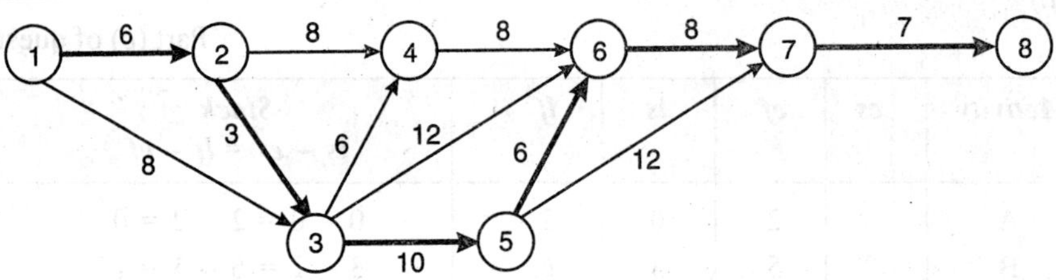

Fig. 5.14 (a)

[**Ans.** Critical path = 1 – 2 – 3 – 5 – 6 – 7 – 8 = 40, Sub-critical = 1 – 3 – 5 – 6 – 7 – 8 = 39]

Activity	*es*	*ef*	*ls*	*lf*	*Stack*
					ls – es = lf – ef
1 – 2	0	6	0	6	0 – 0 = 6 – 6 = 0
1 – 3	0	8	1	9	1 – 0 = 9 – 8 = 1
2 – 3	6	9	6	9	6 – 6 = 9 – 9 = 0
2 – 4	6	14	9	17	9 – 6 = 17 – 14 = 3
3 – 4	9	15	11	17	11 – 9 = 17 – 15 = 2
3 – 5	9	19	9	19	9 – 9 = 19 – 19 = 0
3 – 6	9	21	13	25	13 – 4 = 25 – 21 = 4
4 – 6	15	23	17	25	17 – 15 = 25 – 23 = 2
5 – 6	19	25	19	25	19 – 19 = 25 – 25 = 0
5 – 7	19	31	21	33	21 – 19 = 33 – 31 = 2
6 – 7	25	33	25	33	25 – 25 = 33 – 33 = 0
7 – 8	33	40	33	40	33 – 33 = 40 – 40 = 0

2. *From the following data given, construct the project network and identify the critical path, its duration and total slack of each activity.*

Activity :	*A*	*B*	*C*	*D*	*E*	*F*	*G*	*H*
Duration (weeks) :	1	4	3	2	5	2	2	3
Immediate Predecessor :	–	*A*	*A*	*B*	*(C, D)*	*D*	*F*	*(E, G)*

Solution : **Project network** **(Mark the construction very carefully)**

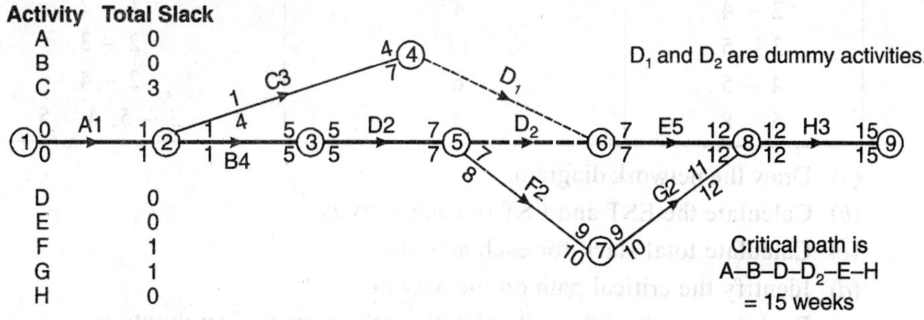

Fig. 5.14 (b)

3. *A project with the following six activities is listed with the normal time period for completion of each activity. Draw the network diagram and find out the time for completion of the project. Also calculate total slack for each activity.*

Activity	*Time duration*	*Immediate predecessor*
A	4	—
B	8	*A*
C	8	*A*
D	5	*B*
E	7	*C*
F	5	*D, E*

Solution :

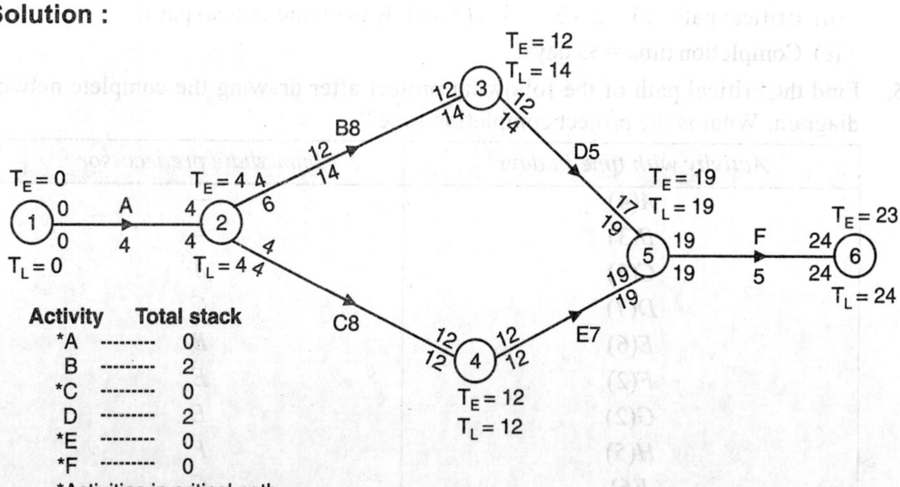

Ans. Time for completion of the project = 24

Fig. 5.15.

4. A small engineering project consists of nine activities.

Activity	*Time duration (days)*	*Immediate predecessor*
1 – 2	6	–
2 – 3	13	1 – 2
1 – 6	5	–
6 – 7	11	1 – 6
7 – 8	15	6 – 7
2 – 4	4	1 – 2
3 – 5	11	2 – 3
4 – 5	6	2 – 4
5 – 8	3	3 – 5, 4 – 5

(*a*) Draw the network diagram.
(*b*) Calculate the EST and LST of each activity.
(*c*) Calculate total slack for each activity.
(*d*) Identify the critical path on the network.
(*e*) Find the length of the critical path or the total project duration.

Solution :

Activity	*EST*	*LST*	*Total Slack = LST – EST*
* 1–2	0	0	0
1–6	0	2	2
* 2–3	6	6	0
2–4	6	20	14
* 3–5	19	19	0
4–5	10	24	14
6–7	5	7	2
* 5–8	30	30	0
7–8	16	18	2

(*d*) Critical path is 1 – 2 – 3 – 5 – 8 (* activities on the critical path)
(*e*) Completion time = 33 days.

5. Find the critical path of the following project after drawing the complete network diagram. What is the project completion time ?

Activity with time in days	*Immediate predecessor*
A(4)	–
B(3)	–
C(5)	*A*
D(7)	*C*
E(6)	*B*
F(2)	*E*
G(2)	*E*
H(5)	*F*
I(6)	*G*
J(4)	*D*,*H*, *I*

(Critical path *BEGIJ* = 21 days)

5.8. PERT COMPUTATION

The network methods discussed so far (i.e., CPM) dealt with deterministic models since estimated activity times are assumed to be the expected values.

Critical path method is an effective tool for project, planning and control when activity times are known with certainty. However, in certain projects like maintenance work, oil digging operations, and research and development work, it seems unrealistic to assume that we can know with certainty the time duration in which the activities can be completed.

Deterministic network methods assume that the expected time is the actual time taken. Probabilistic methods, on the other hand, assume the reverse, more realistic situation, where activity times are represented by a probability distribution. This probability distribution of activity time is based upon three different time estimates made for each activity.

t_0 = the optimistic time is the shortest possible time to complete the activity if all goes well.

There is more than one chance in 100 of completing the activity in this amount of time. It represents an ideal estimate.

I_p = the pessimistic time is the longest time that an activity could take if every thing goes wrong it occurs with a probability of less than 1 percent.

t_m = the most likely time is the estimate of normal time an activity would take. It is the modal time.

From the above three time estimates, the expected time of the activity is computed using following relationship.

$$t_e = \frac{1}{6}(t_0 + t_p + 4\,t_m)$$

The probability distribution of the time involved in performing the activity might be characterized by Beta distribution of the curve shown below. It is assumed that the distribution is unimodal and the peak will represent the value of t_m. PERT assumes that all the job times (t_0, t_p, t_m) are independent and are identically distributed. Hence, by **Central Limit Theorem*** of Statistics the time of completion has a normal distribution with mean t_e and S.D. = σ.

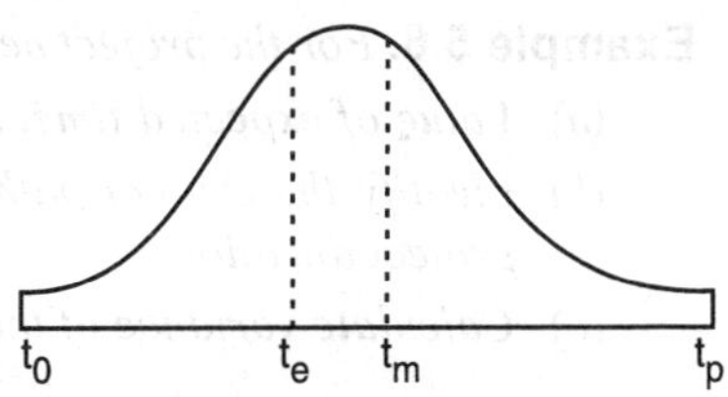

Fig. 5.16. Beta Distribution

There is very little chance that the activity can be completed before t_0 and after t_p. The point t_m is free to move between these two times. Another assumption is that for unimodal distribution, the standard deviation can be estimated roughly as one sixth range, i.e, $\sigma = \dfrac{t_p - t_0}{6}$ and hence the variance = $\left(\dfrac{t_p - t_0}{6}\right)^2$.

---

* **Central Limit Theorem :** Let $X_1, X_2, ..., X_n$ be independent and identically distributed random variables. Then for large N, the sum of random variables $(X_1 + X_2 + ... + X_n)$ is normally with mean $\sum_{i=1}^{N} E(X_i)$ and variance $\sum_{i=1}^{N} V(X_i)$

5.8.1 Probability of Meeting the Scheduled Dates

After identifying the critical path and the occurrence times of all activities, the question arises as to what is the probability that a particular event will complete on or before the scheduled date. The probability that the activity $i-j$ will be completed in time t_e is **fifty per cent**. In the frequency distribution curve the vertical line through t_e will divide the area under the curve into two equal parts. For completing the activity in any other time, say t_r the probability will be,

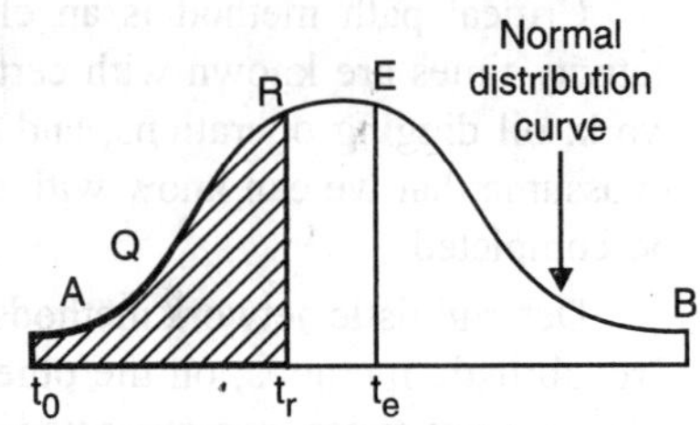

$$\text{Probability} = \frac{\text{Area under AQR}}{\text{Area under AREB}}$$

Fig. 5.17.

The probability depends upon the location of the point t_r. Taking t_e as the reference point, the distance $t_e - t_r$ can be expressed in terms of standard deviation. Since the standard deviation for a normal curve (the above curve as shown is called a normal distribution curve) is 1, and it has an area equal to unity the σ calculated as = $\sqrt{\text{Sum of variances along critical path}}$ is used as a scale factor for calculating the normal deviate (Z).

$$Z = \text{Normal deviate} = \frac{\text{Due date} - \text{Expected date of completion}}{\sqrt{\text{Project variance}}} = \frac{X - \bar{X}}{\sigma}$$

The values of probability for a normal distribution curve corresponding to the different values of normal deviate can be seen from the table "Area under Standard normal Curve".

Note : If there are more than one critical path for a project, the one with higher variance will have to be considered for computation of the chance of its completion.

Example 5.6. *For the project network shown below with times t_o, t_m and t_p, calculate :*

(*a*) *Value of expected time, standard deviation and variance of each activity.*

(*b*) *Identify the critical path and find the length of the critical path or the total project duration.*

(*c*) *Calculate variance of the critical path.*

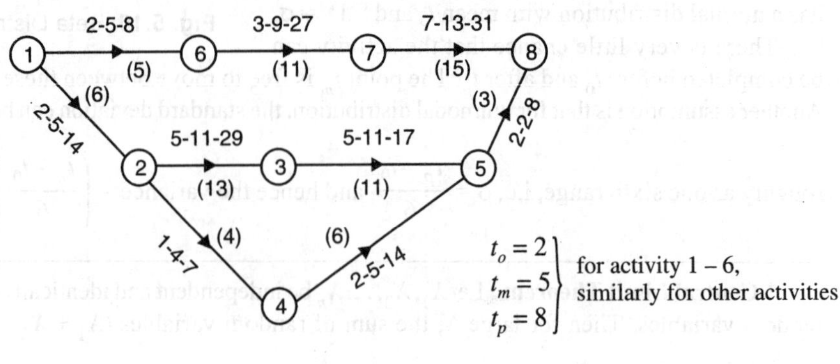

Fig. 5.18.

Solution : (times given with each activity are $t_0 - t_m - t_p$)

(*a*)

Activity	$t_e = \dfrac{t_0 + t_p + 4t_m}{6}$	*Variance* $= \left(\dfrac{t_p - t_0}{6}\right)^2$
1–2	$t_e = \dfrac{2+14+4\times5}{6} = 6$	$V_{1-2} = \left(\dfrac{14-2}{6}\right)^2 = 4$
1 – 6	5	$V_{1-6} = 1$
2 – 3	13	$V_{2-3} = 16$
2 – 4	4	$V_{2-4} = 1$
3 – 5	11	$V_{3-5} = 4$
4 – 5	6	$V_{4-5} = 4$
6 – 7	11	$V_{6-7} = 16$
5 – 8	3	$V_{5-8} = 1$
7 – 8	15	$V_{7-8} = 16$

On the network diagram, t_e has been shown in ( ) bracket.

(*b*) Critical path is 1–2, 2–3, 3–5, 5–8 = 6 + 13 + 11 + 3 = 33 days.

(*c*) Variance of the critical path = Sum of variances of each critical activity

$$= V_{1-2} + V_{2-3} + V_{3-5} + V_{5-8}$$
$$= 4 + 16 + 4 + 1 = 25$$

5.9. DISTINCTION BETWEEN PERT AND CPM

The PERT and CPM models are similar in terms of their basic structure, rationale and mode analysis.

However, there are certain distinctions between PERT and CPM networks which are given below :

(*i*) CPM is activity oriented i.e., CPM network is built on the basis of activities. Also results of various calculations are considered in terms of activities of the project. On the other hand PERT is event oriented.

(*ii*) CPM is a deterministic model i.e., It does not take into account the uncertainties involved in estimation of time for execution of a job or an activity. It completely ignores the probability element of the problem. PERT, however is a probabilistic model. It uses three estimates of activity time, optimistic, pessimistic and most likely; with a view to take into account to uncertainty. Thus, the expected duration of each activity is probabilistic and expected duration indicates that there is fifty per cent probability of getting the job done within that time.

(*iii*) CPM places dual emphasis on time and cost and evaluates the trade-off between project cost and project time. By deploying additional resources, it allows the critical path to manipulate project duration within certain limits so that project duration can be shortened at an optimal cost. On the other hand, PERT is primarily concerned with time. It helps manager to schedule and coordinate various activities so that "the project can be completed on scheduled time."

(*iv*) CPM is commonly used for those projects which are repetitive is nature and where one has prior experience of handling similar projects. PERT is generally used for those projects which time required to complete various activities are not known as priori. Thus, PERT is widely used for planning and scheduling, research and development projects, and CPM is used constructional and business problems taking from space programmes to marriage proposals to horse shows.

MODEL QUESTIONS

1. What is a critical path ? Why is it so important in scheduling and controlling large projects? Can a critical path change during the course of project ? Why ?
2. Define a dummy arrow used in a network. State two purposes for which it is used. Mention four conventions that are used in drawing a network.
3. What is meant by graphing in network analysis ?
4. Explain the following terms :
   (*a*) Total float (*b*) Independent float
   (*c*) Interference float (*d*) Safety float (free float)
5. Distinguish between PERT and CPM.
6. Explain various basic steps in PERT/CPM techniques.
7. Critically comment on assumptions based on which PERT/CPM analysis is done for the projects.
8. Explain the following terms in PERT:
   (*i*) Pessimistic time
   (*ii*) Optimistic time
   (*iii*) Most likely time.
9. The local chapter of an institute is planning a dinner meeting with a nationally known speaker and you are responsible for organizing it. How could PERT/CPM methodology be useful for this situation.
10. Explain how to determine the probability of completing a project in a given time.
11. Explain an AON diagram. How does it differ from the usual arrow diagram ?
12. Write a short note on PERT/CPM network in Operations Research.
13. A small project is composed of seven activities whose time estimates are listed below. Draw the project network diagram and identify all the paths through it. Find the expected time and variance of each activity. What is the expected project length?

Activity	*Optimistic time*	*Most likely time*	*Pessimistic time (weeks)*
1 – 2	1	1	7
1 – 3	1	4	7
1 – 4	2	2	8
2 – 5	1	1	1
3 – 5	2	5	14
4 – 6	3	5	8
5 – 6	3	6	15

Ans. Expected project length = 17 weeks
Critical path = 1 – 3 – 5 – 6 = 17 weeks
Variance of each acvitity = 1, 1, 1, 0, 4, 1, 4 respectively.

CHAPTER 6

SUPPLY CHAIN MANAGEMENT

INSIDE THIS CHAPTER

6.1. INTRODUCTION (MEANING)

The global compoetitiveness means that the customer is supreme. The customers can source his goods and services from anywhere in the world. The objective can be met through proper supply chain management. The evolution of moving materials/products is from the conventional distribution in logistic management and then to the present day supply chain management. Alternatively, one can call the supply chain management as the integrated logistic system. This has emerged as a new buzzword in the corporate world to solve the problems of complexities in the supply chain of the organization.

The early 1990s, saw the beginning of the era of supply chain management which is the integration of all the important business processes. It involves the management and coordination of the movement of products, services and information through the entire supply chain process, from acquiring raw materials through the delivery of the products to supplying final products to the customers. It evolves the interaction and coordination of many organizations and service-industries in the field of manufacturing, selling, transportation, distribution and information systems and management. The process of supply chain management comprises of different phases, namely generation of requirements, sourcing, pricing and post-award activities through the application of IT which includes communication, e-mails, EDI (Electronic Data Interchange) and ERP (Enterprise Resource Planning).

6.2. DEFINITION

According to Menzer, "Supply chain management is defined as the systematic, strategic coordination of the traditional business functions and the tactics across these business functions within a particular company and across businesses within the supply chain, for the purposes of improving the long term performance of the individual companies and the supply chain as a whole".

It is the upstream of the organization's vlaue chain and is responsible for ensuring that the right materials, services and technology are purchased from the right sources at the right time, in the right quality. Supply chain management helps in providing the best customer service in a cost-efficient manner.

6.3. SUPPLY CHAIN MODEL (SCM)(OVERVIEW)

The Fig. 6.1 below shows the supply chain management model representing the coordination of product, which flows across organizational functions and various organizations to achieve competitive advantage and profitability.

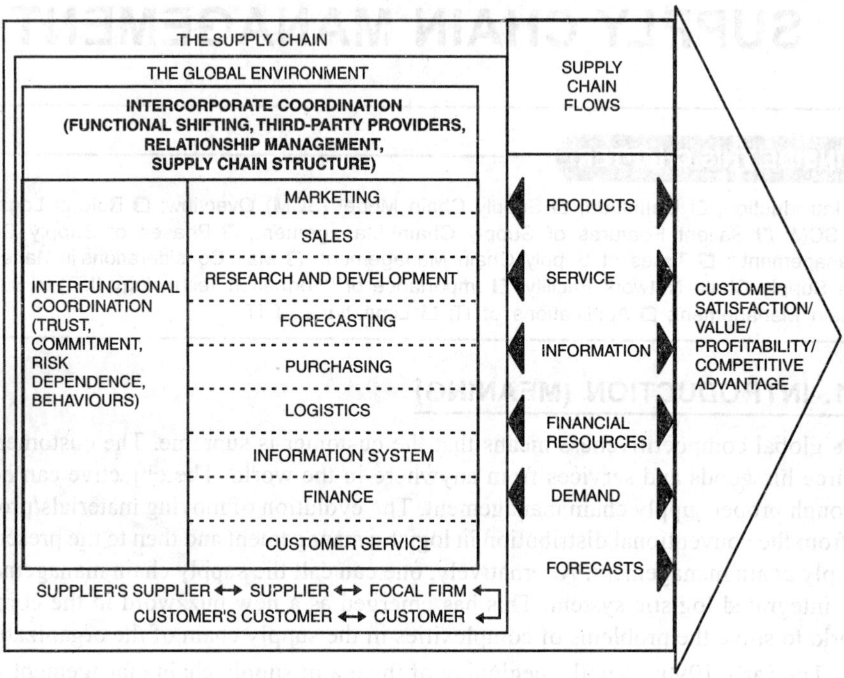

Fig. 6.1. A Supply Chain Management Model

6.4. ROLE OF LOGISTICS IN SCM

Logistics is an essential component of SCM. It involves planning, carrying out the management of goods, services and information from the point of origin to the point of consumption—companies see logistics as a critical blue print of supply chain. The dictionary meaning of logistics is "the branch of military science and operations dealing with the procurement, supply and maintenance of equipment, the movement of personnel, the provision of facilities and the related matters". Logistics is about moving materials or goods from one point to another—logistics being the servant of design, productions and marketing.

Seven major functions of logistics in SCM are the following:

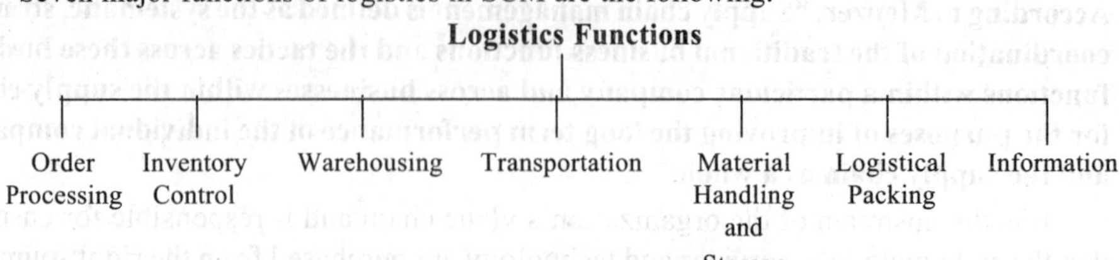

Inbound logistics covers the activities concerned with obtaining materials and their handling, storing and transporting them while the outbound logistics covers the activities concerned with collection, maintenance and distribution to the customers. Other activities such as packing, fulfilling orders warehousing, managing stocks and maintaining the equilibrium between supply and demand.

Logistics management is that part of supply chain management that plans, implements and controls the efficient, effective forward and reverse flow and storage of goods, services and related information between the point of origin and the point of utilization in order to meet customers requests.

6.5. SALIENT FEATURES OF SUPPLY CHAIN MANAGEMENT

The following are the important features of the supply chain management that help in providing the competitive edge to the organization :

***(i)* Single entity :** For a variety of planning and control functions across the supply chain, the responsibility is given to the single or particular group or person. For instance, a group consisting of the representatives from purchase, manufacturing, distribution and sales could be the entity for finalizing the marketing plan, the despatch plan, the production plan and the procurement plan. The single entity helps in reducing the administrative delays in the supply chain.

***(ii)* Inventory perspective :** Traditionally, the inventories have been viewed as an intermediary to reduce the coordination requirements across several activities. Now-a-days, inventory is viewed as an intermediary to be used as the last resort after ensuring proper information sharing and coordination. It is essential for the organization to quantify the level of inventory so as to protect themselves in the uncertain conditions such as stock-out situations. This results in an efficient system, which is more responsive in the long run of business. Some of the key measures that may enable this perspective of inventory in supply chain management are as follows :

(*a*) Improving flexibility

(*b*) Reducing lead-times

(*c*) Reducing uncertainties

(*d*) Improving quality.

***(iii)* Strategic decision-making :** The decisions in the supply chain are viewed as having strategic implications rather than just operational ones. For example, rather than being concerned with just sourcing trucks from the market, the firm may consider long-term contracts with transporters or investing in their infrastructure to suit the firm's purpose. This facilitates smoother and more reliable transport logistics in the long run.

***(iv)* Systems approach :** The supply chain from the vendor to the customer is viewed as a single integrated system rather than many subsystems interfacing with each other. For example, traditionally the efforts may have aimed at developing better interfaces between marketing and production or sourcing and conversion. While this facilitates smooth operational functioning, and issues such as outsourcing, which significantly cut across both sourcing and conversion, may not be considered. The integrated concept of the supply chain may enable such situations.

***(v)* Better performance :** In various activities of the supply chain, it is important to focus on performing the best activity. This has certain implications on the outsourcing and even insourcing business activities and building effective partnerships. When the requirements of logistics are more extensive, then insourcing is the right direction for the organization but now-a-days organizations prefer outsourcing.

***(vi)* Supply chain relationships :** The supply chain concept emphasizes harmonious relationships among the members of the organization such as vendors, channel participants and all third-party logistics service providers. The contributions of supply chain members are significant for the performance of the organization. For instance, the experts in supply chain management suggest a limited number of vendors along their high level of motivation to take appropriate actions for achieving the objectives of the buying firm.

***(vii)* Flexible approach :** Now-a-days organizations design their supply chain for bringing flexibility in the supply chain system. Since business conditions are certain to change, the supply chain configuration must be flexible.

In case the organization does not manage its supply chain, then its delivery chain fails which results in customer dissatisfaction and business loss. Supply chain performs the following functions apart from providing goods and services to customers in time:

(*a*) Minimizing uncertainty in delivery and service.

(*b*) Managing demand with promotion and branding of goods

(*c*) Competing on service by way of value delivery to the customer

(*d*) Reducing lead-times of supply

(*e*) Improving process quality.

(*f*) Initiative at industry level to improve poor infrastructure.

6.6. PHASES OF SUPPLY CHAIN MANAGEMENT

Following are the four phases:

Generation of Requirements : It is a critical activity that results in the identification of the optimal materials and services. It includes the purchases along with the development of the specifications and statements of the work describing these supply chain requirements. The supply chain management must be involved during the generation of requirements to ensure that all the commercial issues such as cost of goods, availability and substitutes receive appropriate consideration.

Sourcing : The objective of sourcing is the identification and selection of the supplier whose cost, qualities, technologies, dependability and services meet the requirements of the firm. The development of supply chain management alliances is a sourcing activity.

Pricing : The objective of pricing is the development of prices that appropriately reward the supplier for its efforts, and that result in the lowest total costs of ownership for the customer firm. The significant role of supply chain management is to determine the price during the pricing phase.

Post-award activities : It ensures that the firm receives the ordered products on time and at a specific price and in the specific quantity. Post-award activities include development of supplier, technique assistance and troubleshooting.

6.6.1. Strategic Supply Management Activities

The development and management of strategic supply chain activities and networks are the most challenging and exciting aspects of the supply chain management. It improves the supply chain relationships so as to attain a sustainable competitive advantage. Supply chain management must focus on the following strategic activities :

(i) Environment monitoring : Supply chain management must understand the markets and monitor the envirormnent to identify the organizational threats and opportunities. These threats and opportunities include material shortages that affect one or more industries, which supply the organization. Shortages affect both, the price and availability of purchased materials and supplies. The firm must take actions to minimize the impact of such shortages by monitoring the changes in the environment.

(ii) Integrated supply strategy : Supply chain management must develop and manage the firm's supply strategy as an integrated series of unrelated strategies. The technology and marketing and production strategies are all the inputs to the integrated supply strategy.

(iii) Commodity strategies : Supply chain management must develop and update sound commodity supply strategies. The following activities must be performed in order to ensure the effectiveness of the firm's commodity strategies :

(*a*) *Strategy updating:* The organization must recognize materials, equipments and services that are strategic in nature, and develop strategic plans for obtaining them. The organization must react quickly by improving and updating its commodity strategies by taking appropriate steps. This also includes **Make/Buy decision.** The product to be manufactured is thouroughly analysed to determine whether it is to be made in the factory or to be purchased in view of the following facts: (*i*) cost comparison for making or buying, (*ii*) whether the available manpower can be utilized for the purpose of working within the plant, (*iii*) whether the existing machines which are idle can be utilized gainfully.

(*b*) *Technology access and control :* The organization must identify the critical or strategic technologies developed by the suppliers with the firm's funds. The organization must take actions to protect the technologies that yield a competitive advantage, and ensure that they are not transferred to competitors.

(*c*) *Supply management organization :* The internal supply chain management of the organization must concentrate on improving the effectiveness and efficiency of the organization.

(*d*) *Risk Management :* It involves taking action for reducing the possibility of any disruption in the supply of products and increase in their price.

(iv) Data Management : Supply chain management must emphasize the collection and application of data, with the aim of facilitating strategic supply planning. The strategic supply planning requires data related to process, which functions within the functional boundaries and company boundaries. The supply chain management alone should not be responsible for planning, controlling and compiling, the supply data must also be considered for analyzing each supplier relationship.

6.7. TYPES OF SUPPLY CHAIN MANAGEMENT

Based on the scope of organization, there are two types of supply chain—intra-organization and inter-organization. The intra-organizational supply chain includes sharing of information with buyers, suppliers and competitors in order to increase sales, whereas in the inter-organization supply chain, information is shared within and outside the organization. Intra-organizational information sharing is done at two levels, data level and business-process level. In data level information sharing, all distribution sources must be properly connected with the supply chain. The data can be recovered from one database such as Enterprise Resource Planning (ERP) to update another database. This approach is easy to implement, as application changes in data transactions are not needed. In business-process-level information sharing, techniques such as Application Programming Interface (API) are applied to data transactions for developing particular integration modules. This approach is challenging because it deals with various difficulties while integrating different data transactions applications.

Generally, intra-organizational information sharing employs middleware technology, while inter-organizational information sharing employs collaboration technology. The middleware technology offers inter-operational services such as business transactions and shared directories among various organizations. The middleware technology also acts as a backbone for using the data into ERP and supply chain management systems. The inter-organization information sharing techniques helps in collaborating different shareholders. According to Amit and Zott, 'Inter-organization collaboration requires sharing critical business information with buyers, suppliers and even rivals and is a concern to those seeking to add value from the increasing scale and scope of electronic applications.' A shift in traditional Inter-Organizational Systems (IOS) is required for achieving efficient supply, development and procurement of goods. The shift in IOS can be motivated by reducing the cost.

Collaboration is an important technique that helps in managing the performance of supply chain that requires maintaining organizational boundaries and cooperative relationships. According to Kanter, Philips and colleagues, 'Collaboration implies shared responsibilities, resources and benefits and the creation of value through common information exchange between organizations.'

6.8. MAIN CONSIDERATIONS IN MANAGING THE SUPPLY CHAIN NETWORK FACILITY

There are four main considerations in managing the supply chain; customer service, facility location, inventory decisions and transportation decisions. All these considerations are inter-related and should be planned properly in order to be cost-efficient. These considerations have the following impact on the supply chain of the organization :

***(i)* Customer service goals :** The services offered to customers affect the supply chain directly. If the organization is offering lower level of customer services, then there is no need of expensive techniques and high level of inventory. On the other hand, higher levels of customer service need expensive techniques and high levels of inventory.

***(ii)* Facility location strategy :** The facility location considers the cost associated in the movement of products from plant to vendor, warehouses and customers. The facility location strategy helps in determining the demand of customers to be served directly from the plant and to the vendors affects the total distribution cost of goods.

***(iii)* Inventory decisions :** These are related to the maintenance of inventory level in the organization.

***(iv)* Transport strategy :** Transportation decisions are related to the selection of transportation mode and the size of shipment. The transportation decisions are affected by the location of warehouse, that is, whether it is near the-market or not.

The decisions related to the customer service level, facility location, inventory and transportation decisions, should be properly taken as they affect the sales, cash in-flow and return of investment of the organization.

6.9. IMPORTANCE OF INFORMATION TECHNOLOGY (I.T.) IN SUPPLY CHAIN MANAGEMENT

The combination of computers and telecommunication offers the fastest means of communication in the business. Now-a-days, the Internet is becoming a reliable link between the suppliers and customers. The biggest advantage of the Internet is that it offers point-to-point information about the market to the customers. Apart from this, IT provides various opportunities to organizations, related to reducing cost and improving profits. Following are the benefits of IT services to organization :

(*a*) It is less capital intensive.

(*b*) It is not location-specific, that is, it can be accessed from anywhere in the world, even from the remote areas.

(*c*) It does not demand expensive infrastructure facilities.

(*d*) It helps in 24-hours communication all over the world that in turn saves manpower.

(*e*) It becomes easy to select the vendors for joint venture, as the whole world gets connected through Internet.

(*f*) It is easy to get the customer feedback quickly and easily through the intemet, which helps in maintaining customer relationship.

6.10. APPLICATIONS OF IT

The various applications of IT help in improving the quality of products and services as it allows you to select the best product and technology from all over the world. There are various applications of IT which include the following :

6.10.1. Communication

The Internet is an important tool for conducting effective business of the organization. The Internet helps in proper flow of communication in the organization and in conducting various activities such as data exchange. Now-a-days, suppliers provide information to customers through Internet rather than distributing catalogues for their products.

6.10.2. e-mail

It builds effective communication throughout the world at any time and at a negligible cost. It is done through the Internet, which is widely accepted medium of delivering business information. Managers prefer e-mail instead of courier for sending the documents because e-mail is time and cost saving.

6.10.3. E-Commerce

It is defined as doing business electronically that includes for sharing of business information by electronic means among suppliers, customers, governmental bodies and other partners in order to conduct and execute transaction in business, administration and consumer activities. Its advantages are:

(*i*) E-C can increase sales and decrease costs.

(*ii*) It increases purchasing opportunities for buyers.

(*iii*) It increases speed and accuracy with which business can exchange information.

(*iv*) It provides buyer with a wide range of choices.

(*v*) E-C can benefit general welfare of society.

(*vi*) Electronic payments are easier to audit and monitor.

(*vii*) E-C enables people to work from home.

(*viii*) It can make products and services available in remote areas.

6.10.4. EDI (Electronic Data Intercharge)

EDI is the exchange of business information through computers in standard formats. In EDI, the business information is organized according to a specified pattern decided by both the organizations. The main advantage of EDI is that it reduces the overall cost and enhances the productivity of the organization. Apart from this, EDI also improves the business relationships with other parties and minimizes the use of paper.

6.10.5. ERP (Enterprise Resource Planning)

ERP is an IT technique that helps in integrating data and processes of the organization into a unified system. ERP uses various components of computer hardware and software techniques for achieving the integration. (See also Sec 4.11, Page 74).

6.11. LIMITATIONS OF IT

Information technology plays an important role in the supply chain by exchanging, storing and utilizing the business data. Despite having several advantages, the IT lacks in the following aspects :

(*a*) Development of Electronic Data Intercharge (EDI) in an organization is costly.

(*b*) Business transactions take place through the Internet, so there is no face-to-face contact between the organizations and suppliers or vendors or customers. Therefore, there is lack of trust between the individuals between whom the transactions take place.

(*c*) The employees, customers and suppliers who are conducting business through phone, fax and meetings, may face difficulties in using the IT techniques because special skills are required for using IT techniques.

MODEL QUESTIONS

1. What do you understand by supply chain management ? Describe briefly.
2. Define a supply chain and explain the concept of supply chain management.
3. What factors are necessary to consider while selecting a strategic plan for creating the supply chain network?
4. What are the four phases of supply chain management ? Describe each of them briefly.
5. What are the different types of supply chain management ? Explain them in short.
6. What do you mean by "Strategic supply management activities ? Discuss them briefly.
7. What are the advantages and limitations of IT services for an organization?
8. Explain the role of logistics in supply chain management.
9. Write short notes on:

   (*a*) E-Commerce

   (*b*) Make/Buy decisions.

□□□

CHAPTER 7

PRODUCTION PLANNING AND CONTROL WITH PRODUCTION SCHEDULING

INSIDE THIS CHAPTER

7.1. INTRODUCTION (NECESSITY AND DEFINITION)

When a product is to be manufactured, the product must be carefully planned to ensure that proper methods are used and every work centre is assigned the amount of work according to its capacity. While the product is being manufactured, the production must be controlled to ensure that the planned output is constantly maintained. Thus, production planning and control (PPC) consists of two different and distinct functions PRODUCTION PLANNING + PRODUCTION CONTROL.

Production planning involves management decisions relating to how much to produce, what materials, parts and tools will be needed, what steps should be followed in the production process, within what time limit the production is to be completed and how much work to be done by each workstation. Production planning is a pre-production activity involving arranging facilities and designing the production system. It is based on sales commitments as to quantity, delivery dates, price and quality, etc.

Production control involves implementation of production plans or schedules by coordinating different activities. It seeks to ensure that production operations and actual performance occur according to planned operations and performance. Production operations are constantly evaluated, guided and directed along the plans formulated by the planning department. Production control endeavours to give the best possible service, consistent with available capacities and at the same time to secure the lowest possible cost.

Thus, PPC may be defined as **"the process of planning production in advance of operations, establishing the exact route of each individual item, part or assembly, setting staring and finishing dates for each important item or assembly and the finished products, and releasing the necessary orders as well as initiating the required follow-up to effectuate the smooth functioning of the enterprise."**

***Production Planning* consists in coordinating the activities of the production activities of the production department with other departments of the business, such as sales, purchasing, material control, personnel, engineering and administrative groups, so that each department may contribute to production activities in the most effective manner.**

7.1.1. Objectives of Production Planning and Control

The major objectives of PPC are the following:

(*a*) To design a system by which production can be achieved to meet promised delivery date with minimum cost and quality standard.

(*b*) To ensure efficient utilization of production facilities.

(*c*) To coordinate the production activities of various departments.

(*d*) To maintain appropriate stock of work-in-process, raw materials and finished products to meet the production requirements.

(*e*) To ensure the production of right product, a right quality at right time.

(*f*) To ensure smooth flow of materials by eliminating production bottle necks if any.

(*g*) To coordinate equipment and labour in most effective and economic way.

(*h*) Establishing goals and checking the performance against set targets.

(*i*) To provide alternative production strategies in case of emergent condition.

(*j*) To find out the magnitude and nature of various inputs for the manufacture of desired output.

7.1.2. Process Planning

A process is any group of actions instrumental to the achievement of the output of an operations system in accordance with a specified measure of effectiveness. When a product of an enterprise is designed, certain specifications are established; physical dimensions, tolerances, standards and quality are set forth. Then, it becomes a matter of deciding the specific details of how to achieve the desired output and this decision is the essence of **Process Planning.**

Process Planning is defined as a function within a manufacturing facility which establishes the processes and process parameters to be used alongwith the machines capable of performing these processes in order to convert a piece-part from its initial form to a finished form that is predetermined on a detailed engineering drawing.

7.1.2.1. Steps Involved in Process Planning

The following steps are involved in process planning for the conversion of design data to work instruction:

(*a*) Quantity of work to be done alongwith product specification.

(*b*) Quality of work to be completed.

(*c*) Availability of equipments, tools and personnels.

(*d*) Sequence in which the operations will be performed on raw material.

(*e*) Details of equipments on which the operations will be performed.

(*f*) Standard time for each operation.

(*g*) When the operations will be performed.

7.1.2.2. Procedure of Process Planning

The following are the procedural steps :

(*i*) Preparation of working drawings.

(*ii*) Deciding to make or buy.

(*iii*) Selection of manufacturing process.

(*iv*) Machine capacity and machines/equipments selection.

(*v*) Selection of materials and bill of materials.

(*vi*) Selection of jigs, fixtures and other attachments.

(*vii*) Operation planning and tooling requirements.

(*viii*) Preparation of documents such as operation sheets or route sheets.

7.2. DEFINITION—PRODUCTION CONTROL

Production control is defined as the design and use of a systematic procedure for establishing plans and controlling all the elements of an activity which includes a follow-up procedure for determining how closely the plan is followed and a means of regulating execution to meet the plan's requirements.

7.2.1. Objectives of Production Control

(*a*) Issuing the necessary orders to the concerned personnel so that the production plans may be put into effect.

(*b*) To arrange necessary items like materials, machines, man power and jigs and fixtures in quantity and quality at a time as planned.

(*c*) To ensure that orders and instructions are followed so that goods with required quality and quantity may be manufactured and delivered at the promised time.

7.2.2. Fields of Production Control

(*i*) Control of activities

(*ii*) Control of materials movement

(*iii*) Control of tools availability

(*iv*) Control of due date

(*v*) Control of quality produced

(*vi*) Control of quantity produced

(*vii*) Control of labour efficiency.

7.3. FUNCTIONS (COMPONENTS) OF PRODUCTION PLANNING AND CONTROL

These are:

(*i*) Planning

(*ii*) Routing

(*iii*) Loading

(*iv*) Scheduling

(*v*) Dispatching
(*vi*) Progressing or follow up
(*vii*) Inspection

These functions can be divided in planning, action and control phases, and are dependent on each other and have very closed inter-relationship.

(A) Planning Phase :

(*a*) *Preplanning :*

- Forecasting
- Estimating the work requirement in future
- Estimating the cost of work
- Product design. This includes the preparation of specification of new products.

(*b*) *Action Planning :*

- Process planning and routing : Preparing work details
- Material requirement planning
- Tool requirement planning
- Scheduling
- Men and machine loading : determining men and machine requirements and their control

(B) Action Phase :

- Dispatching : Starting the work
- Follow-up, *i.e.*, carry out work as per action plan
- Initiate material purchase, receipt and storage, etc.

(C) Control Phase :

(*a*) *Progress reporting*

- Collection of data regarding progress of work and inventory, etc.
- Interpreting the data.

(D) Corrective Action :

- Expediting *i.e.,* making correction in the work swiftly
- Replanning *i.e.,* making correction in the plan and then implement.

7.4. PLANNING

Planning is the act of making schemes for various activities.

Before starting any work, it is necessary to plan the things properly for getting the best result. Similarly for production, planning in advance is very necessary and if there is no planning, then how shall we control ? It is the planning department, which decides about each element of the job in anticipation that what work will be done, where, how and when it will be done. The functions of planning are grouped as under :

(*i*) Investigation about the complete details and requirements of the product to be manufactured.
(*ii*) Pre-determination of future achievements.
(*iii*) Planning the design of product to be manufactured.

(*iv*) Planning about the quantity of materials which are to be consumed.

(*v*) Planning about the standard of quality to be manufactured.

(A complete list of all parts required for manufacture of a specified products that is bill of materials to be made).

(*vi*) Planning about the sequence of operations.

(*vii*) Planning about the capacity of equipments.

(*viii*) Planning about internal transportation.

7.5. ROUTING

Routing may be defined as the selection of path which each part of the product will follow while being transformed from raw materials to the finished product.

7.5.1. Functions of Routing

(*a*) Studying the product, determination of the alternative methods of processing and selecting the best method for manufacturing.

(*b*) Determining the equipment, regular or special where the processing will be done.

(*c*) Determination of the capacity of the machine available for processing.

Route Sheet

Manufacturing Order No.		Date Of Completion.................
Description Of Part		Part Symbol ....
Number Of Pieces On Order Or In Lots	Spoilage Allowance	Total To Make ....

Material Symbol :	Material Description		Quality Of Material Allotted ....
Purchase Order No.	Due Date	Stores Requisition No..........	Move...........................

Operation No.	Tool		Operation	Machine	Time Allowance	Due Dates		Move Order	Pieces Good
	Making Order					Begin	Finish		
1			Turning						
2			Screw Cutting						
3			Grinding						

Fig. 7.1. Route Sheet

(*d*) Determination of the sequence of operations to be carried out for completing the product.

(*e*) Determination of speeds at which operations will be performed.

(*f*) Determination of time required for each operation. Total time of operation includes machine setup time, operating time, process operation allowances for heat treating, metal finishing, etc. bench work time, inspection time and time for moving the work-in-process.

(*g*) Preparation of *rout sheets.* Where a completed product consists of a number of separate parts assembled together, a route sheet is prepared for each of such parts. A route sheet (Fig. 7.1) shows the name of the part, manufacturing order number, materials to be used and operations to be performed in sequence. There are columns for recording against each operation the department where the operation will be performed, the machine on which it will be done, tools to be used, time taken and inspection report as to number of units passed as good and number of units rejected.

(*h*) Grouping or route sheets for sub-assemblies and major assemblies. The route sheets of parts required for a sub-assembly or assembly should be grouped and production of the parts so timed that all of them are completed, when assembly work is commenced.

(*i*) Preparation of *works orders,* also called *manufacturing* or *production orders,* for authorizing production according to details continued in the works order (Fig. 7.2).

(*j*) Adapting procedures to the dispatching system in use in the plant.

The first six of these functions overlap with process engineering. When a process engineering section independently deals with these functions, the last four functions are dealt with by the production planning and control department.

Works Order

No. 20.	PRODUCT XYZ		DATE 5.1.09
Date To Commence	Number Required	Date To Complete	Actual Date of Completion
15.1.09	500	30.1.09	
PART NUMBER	PART NAME	Materials Required	
		CODE NO.	NAME
1	P-1	C-1	M-1
2	P-2	C-2	M-2
3	P-3	C-3	M-3
4	P-4	C-4	M-4
5	P-5	C-5	M-5
6	P-6	C-6	M-6
			Signed by ....

Fig. 7.2. Works Order

7.6. LOADING—DEFINITION AND CONCEPT

Loading may be defined as the assignment of work to a facility. The facility may be people, equipment, work groups, or an entire plant. Therefore, machine loading is the process of compiling the load or quantity of work assigned to each machine for a given period of time. Machine loading charts can be used to show the availability of machine and production manager can plan it to use for full capacity. If the machine is under loaded, he can plan to put more work, while the machine is overloaded, he can either plan for overtime (for temporary overloading) or providing additional machine and manpower (for permanent overloading).

7.6.1. Main Objectives of Machine Loading

(*a*) To engage the machine continuously at work.

(*b*) To assign dates for producing each part of a product so that the total orders can be processed in the shortest possible time.

Further there are two terms generally encountered with the term machines loading:

(*i*) **Overloading.** This may be temporary or permanent. With temporary overloading, the loading can be reduced by over-time work or by diverting the jobs to other machines, if available. While permanent overloading can be met by additional equipments, change in the processing method or by procuring the component by direct purchasing or sub-contracting.

(*ii*) **Underloading.** This is due to less work. In mass manufacturing, underloading can be avoided by producing similar components for stock. Many times underloading is due to inefficient management and wrong planning.

7.6.2. Types of Loading

There are two types of loads. First the load consists of open orders that is existing orders from customers assigned to work centre and secondly the planned orders that is prospective orders from customers.

7.6.2.1. Progress Control Charts

After the work has been scheduled, a progress control system is established. For this, following control charts are necessary :

These charts are used whenever there is discrepancy, these indicate the need for investigation and possible correction.

1. Project Layout Sheet

This is prepared in advance and shows relative importance of process to be done one after the other, *i.e.*, the sequence in which the part should be processed. A typical project layout sheet is shown in Fig. 7.3.

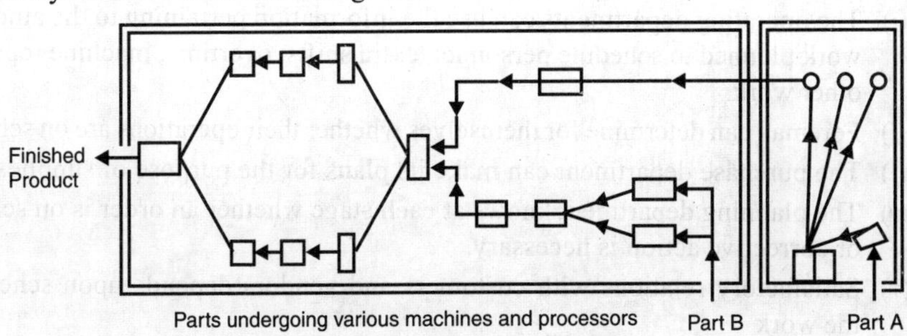

Fig. 7.3. Project Layout Sheet

2. The Load Chart (Fig. 7.4)

This shows the work ahead of each machine with respect of time, i.e., in weeks or days. This may be prepared for different departments and also for different machines.

3. Man-Machine Chart

This chart describes graphically the activity of man and machine he is attending against a time scale. When a shop has a number of operators and machines, studies can be made to find what number of machines each worker can operate or how many workers can work on each machine. (Fig. 7.4)

During the machining cycle of many semi-automatic machines, the operator has time in which he can run one or more other semi-automatic machines. In this, such charts enable the management to minimise the idle time of the workers and to use the machines to their full capacity.

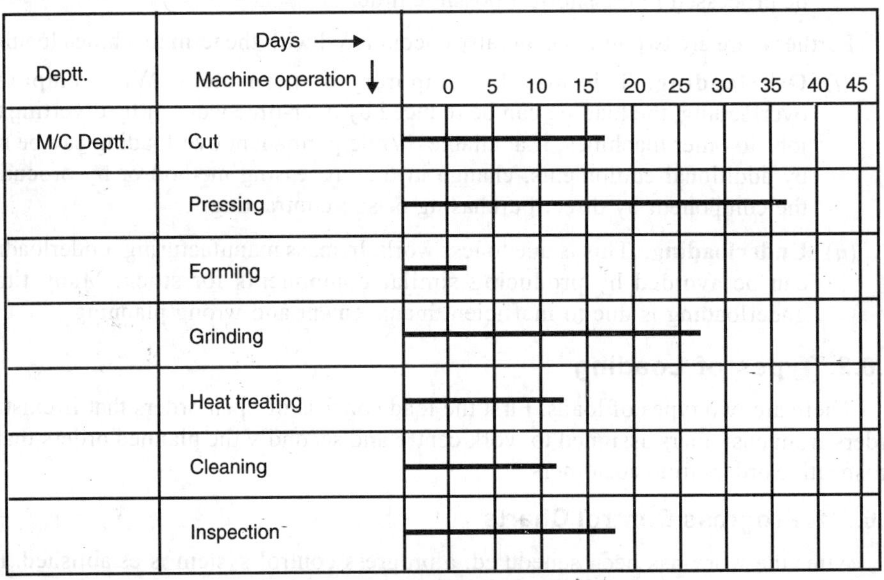

Fig. 7.4. Machine Loading Chart

7.7. SCHEDULING (PRODUCTION SCHEDULING)

Production scheduling may be defined as the fixation of time and date of each operation as well as it determines the sequence of operations to be followed.

7.7.1. Purpose of Scheduling

(*i*) The operating departments can use the information pertaining to the amount of work planned to schedule personnel, extra shifts, overtime, machine repair and other work.

(*ii*) Foreman can determine for themselves whether their operations are on schedule.

(*iii*) The purchase department can make its plans for the purpose of supplies.

(*iv*) The planning department knows at each stage whether an order is on schedule or corrective action is necessary.

(*v*) Satisfactory relations with customers and vendors depends upon scheduling the work.

7.7.2. Types of Scheduling

Scheduling are of the following types : Master Production Schedule and Shop Schedule.

7.7.2.1. Master Production Schedule (MPS)

Scheduling normally starts with the master schedule. A master schedule resembles central office which possesses information about all the orders in hand. Fig. 7.4(*a*) shows a master schedule for a foundry shop which is a weekly breakdown of the production requirements. The total capacity of any week is of 100 hrs. (say) of work in the foundry shop.

Master Schedule for the Foundry Shop Maximum production: 100 hrs. Minimum production : 8 hrs.			
Week 1	Week 2	Week 3	Week 4
15	18	20	15
25	25	12	10
20	28	32	
35			

Fig. 7.4 (a) Master Schedule for a Foundry Shop

As the orders are received, depending upon their delivery dates, they are marked on the master schedule. When the shop capacity is full for the present week the newly acquired orders are carried over to the next week and so on. A master schedule is thus updated continuously, if depicts a running total of the production requirements and shows the work ahead–yet to be completed. Master schedule is actually the basis for all subsequent scheduling techniques.

After framing the overall picture of production requirements through a Master Schedule Chart, the detailed schedules are thought of and made for each component, and sub-assemblies so that all parts are available at the time of assembly. There are a number of visual aids and techniques which aid in detailed scheduling. The techniques to be employed for scheduling purposes depends upon the type of production (intermittent or continuous). MPS finds applications in big firms, for the purpose of loading the entire plant, in R&D organisations and for the overall planning in foundry shops, computer centres and repair shops, etc.

Advantages: It is simple and easy to understand and it involves less cost to make it and to maintain. A certain percentage of total weekly capacity can be allocated for rush orders.

7.7.2.2. Shop or Manufacturing Schedule

After preparing master schedule, shop or manufacturing schedules are prepared. It assigns a definite period of time to a particular shop for manufacturing products in required quantity. It shows how much products are to be prepared and on what day, week or month, etc.

7.8. TECHNIQUE OF SCHEDULING

Some common scheduling techniques are the following :

(*i*) Flow Process Chart

(*ii*) Gantt Chart

(*iii*) Priority Rule in Sequencing and Sequencing in jobs on one, two and three machines. (Johnson's Algorithm)

7.8.1. Flow Process Chart

This is a graphic representation of operation, transportation, inspection, delays and shortages occurring during production. It also gives information regarding distance moved and time required for transportation, delays and inspection etc. The flow process chart for a gear box casting is shown below :

Name of part : Gear Box Casting	O. operation → Transportation	Job No. ....
Drawing No..........	**D.** Delay Δ Stores □ Inspection	Part No. ....
Date of Delivery.........		

Symbol	*Description*	*Distance moved*	*Time in Minutes* *Delay* ⊙	→	D	□
⊙ → □ D Δ	Pig Iron, Coal and Coke in stores	5 m	–	–	–	–
⊙ → □ D Δ	Charge melted	–	3	–	–	–
⊙ → □ D Δ	Awaited melting	–	–	–	4	–
⊙ → □ D Δ	Molten Metal to Moulding	6 m	–	3	–	–
⊙ → □ D Δ	Pouring	–	2	–	–	–
⊙ → □ D Δ	Cooling	10 m	–	–	10	–
⊙ → □ D Δ	Transportation Fitting shop	–	–	5	–	–
⊙ → □ D Δ	Fitting	–	6	–	–	–
⊙ → □ D Δ	Inspection	–	–	–	–	2
⊙ → □ D Δ	Transportation to casting store	10 m	–	5	–	–
⊙ → □ D Δ	Storing	–	–	–	–	–

Fig. 7.5. Flow Process Chart

7.8.2. Gantt Chart

Designed by Henry L. Gantt around 1910, the Gantt chart is used extensively for production control. Time is marked along the horizontal axis and orders in hand in vertical axis. The amount of work planned or scheduled is shown by the firm lines and the machine on which the orders will be processed is marked on the line. The actual progress of various orders is shown in dotted lines (see Fig. 7.6). The cursor placed at today's date indicates the order A.372 is going as per scheduled. Work on order B.260 started one work before the schedule (starting) date and it is about 70% complete which otherwise would have been only 50% complete as per the plan and thus order, B.260 is ahead of the schedule. Order C.300 which started on the scheduled date, has gone delayed by about one week due to some reasons.

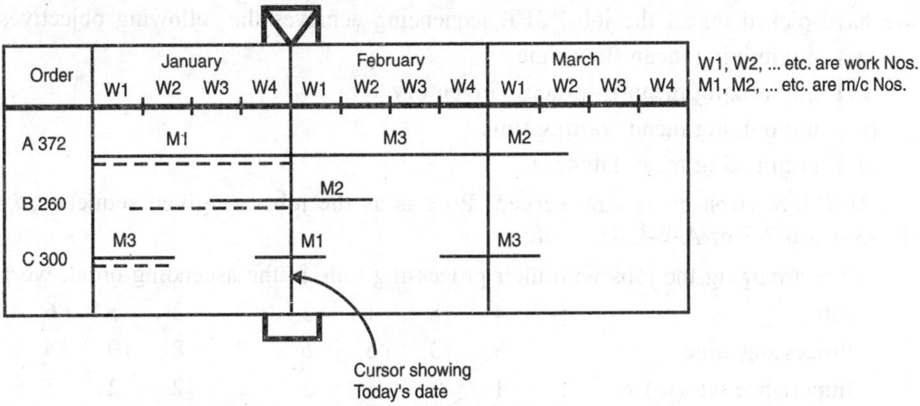

Fig. 7.6. Gantt Chart.

7.8.3. A. Priority Rule in Sequencing *n* Jobs on 1 Machine (Johnson's Algorithm)

We are often facing the problem of sequencing a number of jobs on one machine in practice, for example, a number of patients to be treated by a doctor, a number of cars to be surviced at a service station and a number of broken-down machines to be repaired by a service mechanic, etc. Each job governed by its unique technological requirement demands processing on machine in a certain order in order to minimize mean flow time, mean in-process inventory, and mean waiting time, etc. We shall discuss here the above cases to ascertain as to what is the minimum time for a particular sequencing order to achieve our objective. In this regard we shall deal with the priority rules on sequencing jobs.

Priority Rules in Sequencing Jobs. Some commonly specified priority rules for sequencing jobs are the following:

(*i*) **FC FS :** Select the job on a First Come First Served basis.

(*ii*) **SPT :** Select the job with the Shortest Processing Time.

(*iii*) **EDD :** Select the job with the Earliest Due Date.

(*iv*) **SLACK :** Select the job with the minimum static slack (due date minus time of arrival at machine centre).

(*v*) **RANDOM :** Select the job in random order.

The SPT rule consistently tends to give the lowest mean flow time even for large shops - a property that has been demonstrated in a simple n job × 1 machine case. With the SPT rule, however, the variance of flow times tends to be large. In this regard other rules such as FCFS performs better.

The following example would explain the rules in a better way.

Example 7.1. Consider the 8 jobs with processing times, due dates, and importance weights as shown below:

Job (i)	:	**1**	**2**	**3**	**4**	**5**	**6**	**7**	**8**
Processing time (t_i)	:	**5**	**8**	**6**	**3**	**10**	**14**	**7**	**3**
Due date (d_i)	:	**15**	**10**	**15**	**25**	**20**	**40**	**45**	**50**
Importance weight (w_i)	:	**1**	**2**	**3**	**1**	**2**	**3**	**2**	**1**

1. Shortest Processing Time (SPT) Rule : Here, the job with the least processing time is picked up first, followed by the one with the next smallest time and so on till we have picked up all the jobs. SPT sequencing achieves the following objectives.

(*i*) minimizing mean flow time

(*ii*) minimizing mean in-process inventory

(*iii*) minimizing mean waiting time

(*iv*) minimizing mean lateness.

[F.C.F.S. (first come first served) Rule is as the jobs are given sequence, *i.e.*, 1-2-3-4-5-6-7-8 or A-B-C-D...., etc.]

Now arranging the jobs with their processing time in the ascending order, we get

Job	:	4	8	1	3	7	2	5	6
Processing time	:	3	3	5	6	7	8	10	14
Importance wt. (w_i)	:	1	1	1	3	2	2	2	3

The SPT sequence is 4-8-1-3-7-2-5-6 resulting in completion of these jobs at times 3, 6, 11, 17, 24, 32, 42, and 56 respectively. (For tie, we may take any sequence. For example, t_i for job 4 = t_i for job 8 = 3 we may take first job 4 and then job 8 or the reverse).

$$\therefore \quad \text{The mean flow time} = \frac{3+6+11+17+24+32+42+56}{8}$$

$$= \frac{191}{8} = 23.875 \text{ units} = 23.9 \text{ units}$$

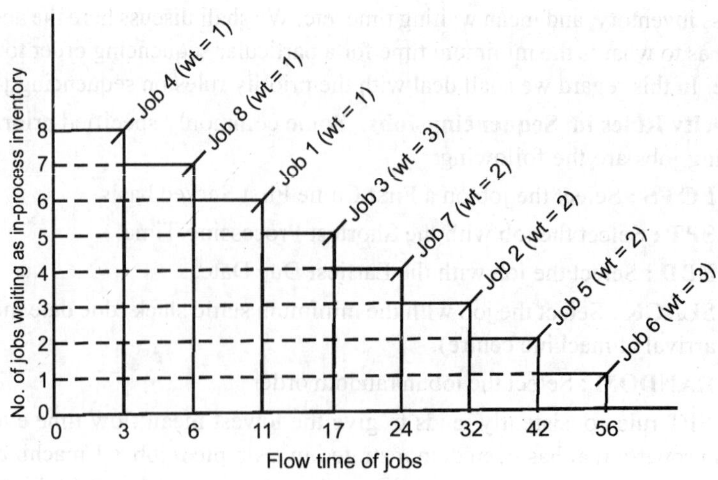

Fig. 7.7.

8 no. of tasks waiting as in-process inventory during time 0–3 (see graph above)	
7 no. of tasks waiting during	3–6
6 no. of tasks waiting during	6–11
5 no. of tasks waiting during	11–17
4 no. of tasks waiting during	17–24
3 no. of tasks waiting during	24–32
2 no. of tasks waiting during	32–42
1 no. of tasks waiting during	42–56

$$\text{Weighted mean flow time} = \frac{1\times3+1\times6+1\times11+3\times17+2\times24+2\times32+2\times42+3\times56}{1+1+1+3+2+2+2+3}$$

$$= \frac{435}{15} = 29$$

Thus, the average in process inventory

$$= \frac{(8\times3)+(7\times3)+(6\times5)+(5\times6)+(4\times7)+(3\times8)+(2\times10)+(1\times14)}{56}$$

$$= \frac{191}{56} = 3.41 \text{ jobs.} \qquad [56 = \text{Total flow time of jobs}]$$

Waiting time for each job

Job number 4	—	waiting time	= 0
Job number 8	—	waiting time	= 3
Job number 1	—	waiting time	= 6
Job number 3	—	waiting time	= 11
Job number 7	—	waiting time	= 17
Job number 2	—	waiting time	= 24
Job number 5	—	waiting time	= 32
Job number 6	—	waiting time	= 42
		Total	= 135

Mean waiting time under SPT rule $= \frac{135}{8} = 16.875$ units.

Lateness of each job.

Lateness = Completion time of job i - due date of job i

Job no. 4, lateness = 3–25 = – 22
Job no. 8, lateness = 6–50 = – 44
Job no. 1, lateness = 11–15 = – 4
Job no. 3, lateness = 17–15 = + 2
Job no. 7, lateness = 24–45 = – 21
Job no. 2, lateness = 32–10 = + 22
Job no. 5, lateness = 42–20 = + 22
Job no. 6, lateness = 56–40 = + 16 (Positive lateness is tardiness)

$$\therefore \text{ Mean lateness} = \frac{-22-44-4+2-21+22+22+16}{8} = -\frac{29}{8} = -3.625 \text{ units.}$$

$$\text{Mean tardiness} = \frac{0+0+0+2+0+22+22+16}{8} = \frac{62}{8} = 7.8$$

(We have taken into account only tardy jobs)

No. of tasks actually late = 4 (Job no. 3, 2, 5, 6) = No. of tardy jobs.

Maximum lateness = 22 units (Job 5 and job 2 each)

2. Weighted Shortest Processing Time (WSPT) Rule : If the importance weights w_i are to be considered WSPT rule could be used to minimise the weighted mean flow time to yield the sequence 3 – 4 – 8 – 7 – 2 – 6 – 1 – 5 (shown below how).

Table 7.1.

Task (i)	*Processing time (t_i)*	*Importance wt (w_i)*	t_i/w_i
1	5	1	5/1 = 5 (VII)
2	8	2	8/2 = 4 (V)
3	6	3	6/3 = 2 (I)
4	3	1	3/1 = 3 (II)
5	10	2	10/2 = 5 (VIII)
6	14	3	14/3 = 4.7 (VI)
7	7	2	7/2 = 3.5 (IV)
8	3	1	3/1 = 3 (III)

The respective flow time of jobs in this sequence are 6, 9 (= 6 + 3), 12 (= 9 + 3), 19 (= 12 + 7), 27 (= 19 + 8), 41 (= 27 + 14), 46 (= 41 + 5), 56 (= 46 + 10)

$$\text{Mean flow time} = \frac{6+9+12+19+27+41+46+56}{8} = \frac{216}{8} = 27 \text{ units}$$

Weighted mean time flow

$$= \frac{(3\times6)+(1\times9)+(1\times12)+(2\times19)+(2\times27)+(3\times41)+(1\times46)+(2\times56)}{3+1+1+2+2+3+1+2}$$

$$= \frac{412}{15} = 27.47 \text{ units} \simeq 27.5 \text{ units}$$

Similarly, we can calculate the following :

Mean lateness = – 0.5

Mean tardiness = 10.6

No. of tardy jobs = 4 (job no. 2, 6, 1, 5)

Maximum tardiness = 36 units

3. Earliest Due Dates (EDD) Rule : Here, the jobs are sequenced in the order of non-decreasing due dates of jobs. The EDD sequence is 2(10 units of time), – 1(15 units) – 3(15 units) – 5(20 units) – 4(25 units) – 6(40 units) – 7(45 units) – 8(50 units). The sequence 2 – 1 – 3 – 5 – 4 – 6 – 7 – 8 yields the completion time of these job as 8, 13 (= 8 + 5), 19 (= 13 + 6), 29 (= 19 + 10), 32 (= 29 + 3), 46 (= 32 + 14), 53 (= 46 + 7), 56 (= 53 + 3). The respective lateness value = – 2 (= 8 – 10), – 2 (= 13 – 15), + 4 (= 19 – 15), + 9 (29 – 20), + 7 (= 32 – 25), + 6 (= 46 – 40), + 8 (= 53 – 45), + 6 (56 – 50). Mean flow time = 32 and weighted mean flow time = 31.7.

$$\therefore \text{Mean lateness} = \frac{-2-2+4+9+7+6+8+6}{8} = \frac{36}{8} = +4.5 \text{ units.}$$

$$\text{Mean tardiness} = \frac{0+0+4+9+7+6+8+6}{8} = \frac{40}{8} = +5 \text{ units.}$$

Maximum lateness = 9 units compared to 22 in SPT.

No. of late jobs = 6 (Job No. 3, 5, 4, 6, 7, 8) = No. of tardy jobs.

The objective of EDD rule is (*i*) to minimise the maximum job lateness. Unfortunately EDD rule tends to make more tardy jobs and at times increases the mean tardiness.

Table 7.2. Comparison of three Scheduling Rules

Rule	*Objective (to minimise)*	*Mean flow time*	*Weighted mean*	*Mean lateness flow time*	*Maximum tardiness*	*No. of tardy jobs*	*Mean tardiness job*
SPT	(*i*) mean flow time (*ii*) mean in-process inventory (*iii*) mean waiting time (*iv*) mean lateness	23.9	29.0	–3.6	22	4	7.8
WSPT	Weighted mean flow time	27.0	27.5	–0.5	36	4	10.6
EDD	Maximum job lateness	32.0	31.7	4.5	9	6	5.0

B. Sequence *n* Jobs on 2 Machines (Johnson's Algorithm)

The problem can be described as (*i*) only two machines *A* and *B* are involved, (*ii*) each job is processed in order *AB*, (*iii*) the exact or expected processing times A_1, A_2, ... A_n and B_1, B_2, ... B_n are known. The problem is to minimise time *T*, the elapsed time from the start of the first job and to the completion of the last job.

It can be shown that the sequences that minimise *T* are the same for both machines. **The method of solving described here (without proof) is due to S.M. Johnson (1954).**

Step 1. Select the smallest processing time occurring in the list A_1, A_2, ..., A_n; B_1, B_2, ... B_n. If there is a tie either of the smallest processing time, should be selected. Then go to step 2.

Step 2.

(*i*) If the least processing time is A_r, selected the *r*th job first, if it is B_s, do the *s*th job last (as the given order is *AB*)

(*ii*) If the tie for the minimum occurs among A_i's, select the job corresponding to the min. of B_i's and process it first of all. For example, let $\begin{pmatrix} A_2 = 3, & A_5 = 3 \\ B_2 = 5, & B_5 = 7 \end{pmatrix}$ then select the job A_2 obviously and process it first of all.

(*iii*) If the tie for the minimum occurs among B_i's select the job corresponding to the minimum of A_i's and process it in the last. For example, let $\begin{pmatrix} A_3 = 4, & A_6 = 3 \\ B_3 = 2, & B_6 = 2 \end{pmatrix}$, then select the job B_6 obviously and process it in the last.

(*iv*) If a set of double ties occur, *i.e.*, say $\begin{pmatrix} A_3 = 5, & A_7 = 5 \\ B_3 = 8, & B_7 = 8 \end{pmatrix}$, then choose arbitrarily.

Job 3 may be first job and job 7 next or the reverse.

Step 3. There are now (*n* – 1) jobs left to be ordered. Apply Step 1 and Step 2 to the reduced set processing times obtained by deleting the two machines' processing times corresponding to the jobs already assigned.

Step 4. Continue in this manner until all the jobs have been ordered. The resulting order will minimise *T*.

Example 7.2. Six jobs go first on machine I and then over II. The order of the completion of the jobs has no significance. The following table gives the machine times in hours for six jobs and the two machines:

Job No. (*i*)	:	**1**	**2**	**3**	**4**	**5**	**6**
Time of machine I (A_i)	:	**5**	**9**	**4**	**7**	**8**	**6**
Time of machine II (B_i)	:	**7**	**4**	**8**	**3**	**9**	**5**

Find the sequence of the jobs that minimises the total elasped time to complete the jobs. Find the minimise time by using Gantt's chart or by other method.

Solution. Out of $A_1, A_2, A_3, A_4, A_5, A_6$ and $B_1, B_2, B_3, B_4, B_5, B_6$, the smallest time is B_4 (*i.e.,* time for job 4 on machine II). Therefore we shall do the job 4 in the last and list the element as follows :

					I
					4

(I denotes the order of choice)

Now of the remaining jobs

i	:	1	2	3	5	6
A_i	:	5	9	4	8	6
B_i	:	7	4	8	9	5

Minimum $(A_i, B_i) = A_3 = B_2 = 4$ $\therefore$ Job A_3 to be done on I and job B_2 goes to next to last job B_4 on machine I.

II				II	I
3				2	4

Now the remaining jobs

i	:	1	5	6
A_i	:	5	8	6
B_i	:	7	9	5

Minimum $(A_1, B_1) = A_1 = B_6 = 5$ $\therefore$ Job 1 to m/c I next to job 3 and job 6 to m/c I next to job 2.

II	III		IV	III	II	I
3	1		6	2	4

Now only one space is left vacant above and it is to be filled obviously by job 5. That completes the sequence as below:

II	III	IV	III	II	I
3	1	5	6	2	4

Now to calculate minimum elasped time

Job	***Machine I***		***Machine II***		***Idle time***
	Time in	***Time out***	***Time in***	***Time out***	***of Machine II***
3	0	4	4	12	4
1	4	9	12	19	—
5	9	17	19	28	—
6	17	23	28	33	—
2	23	32	33	37	—
4	32	39	39	42	2

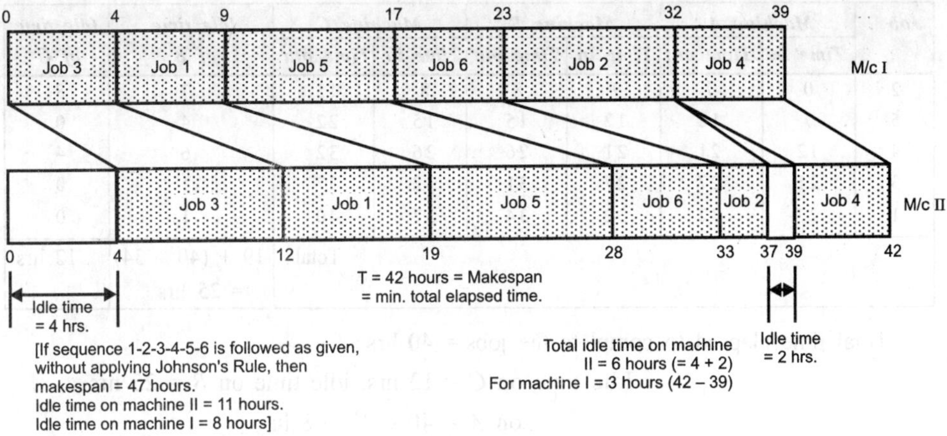

Fig. 7.8. Gantt Chart

C. Sequencing *n* Jobs on 3 Machines (Johnson's Algorithm)

Johnson's rule can be extended to the sequencing of *n* jobs on 3 machines provided either or both the following conditions hold good (if neither of conditions holds good, the method fails):

Condition (*i*) : Min. time on m/c A ≥Max. time on machine B

Condition (*ii*) : Min. time on m/c C ≥ Max. time on machine B

The method is to replace the problem with an equivalent problem involving *n* jobs on 2 machines. If G and H denote two machines (fictitious), then the processing time on G = Times of $A_i + B_i$ and processing time on H = Times of $B_i + C_i$ once the sequencing is completed, the idle times on the 3 m/cs can be calculated separately. The example below will illustrate the application.

Example 7.3. We have five jobs each of which must go through the m/cs *A, B, C* in the order $A \rightarrow B \rightarrow C$.

Processing time in hrs

Job No. (*i*)	**1**	**2**	**3**	**4**	**5**
Machine *A* (A_i)	**5**	**7**	**6**	**9**	**5**
Machine *B* (B_i)	**2**	**1**	**4**	**5**	**3**
Machine *C* (C_i)	**3**	**7**	**5**	**6**	**7**

Determine the sequence to minimize the total elapsed time to complete the jobs and also the idle times on each m/c.

Solution. Min. $A_i = 5$, Max. $B_i = 5$, Min. $C_i = 3$

Since the first condition is satisfied, we can solve the problem by Johnson's rule. We take two fictitious machines G and H such that $G_i = A_i + B_i$ and $H_i = B_i + C_i$.

Job No.	:	1	2	3	4	5
$G_i = A_i + B_i$	:	7	8	10	14	8
$H_i = B_i + C_i$	:	5	8	9	11	10

Sequence optimal by Johnson's rule is 2 – 5 – 4 – 3 – 1

Job	Machine A		Machine B		Machine C		Idle time of B	Idle time of C
	Time in	Time out	Time in	Time out	Time in	Time out		
2	0	7	7	8	8	15	7	8
5	7	12	12	15	15	22	4	0
4	12	21	21	26	26	32	6	4
3	21	27	27	31	32	37	1	0
1	27	32	32	34	37	40	1	0
						Total	19 + (40 – 34) = 25 hrs	12 hrs

Total time elapsed to complete the jobs = 40 hrs

Idle time on C = 12 hrs, idle time on B = 25 hrs

on $A = 40 - 32 = 8$ hrs

The method can further be extended for more than 3 machines, *i.e.*, n jobs X m machines.

Advantages of Sequencing

1. It minimises: (*i*) mean flow time of jobs
   (*ii*) mean in-process inventory
   (*iii*) mean waiting time of jobs
   (*iv*) mean lateness of jobs
   } for a number of jobs on one machine
2. It reduces the number of tardy jobs.
3. It reduces total operation time of different jobs on a number of machines.
4. It helps production manager for advance planning to have a check over the production rate.

7.9. DISPATCHING

Dispatching is the routine of setting productive activities in motion through the release of orders and instructions, in accordance with previously planned times and sequences, embodied in route sheets and schedule charts.

Dispatcher transmits orders to various shops which determines by whom the job shall bve done and it coordinates production. It creates a direct link between production and sales.

7.9.1. Functions of Dispatching

(*i*) To check the availability of input materials and ensure the movement of material from store to first process and then from process to process.

(*ii*) To ensure the availability of all production and inspection aids.

(*iii*) To obtain the requisite drawings, specifications and material lists.

(*iv*) To assign the work to appropriate machine, workplace and men.

(*v*) The issue of job orders authorising operations in accordance with dates and times previously planned and entered on load charts and route sheets.

(*vi*) The issue of time tickets, instruction cards and other required items to the workers who are to perform the various activities.

(*vii*) The issue of inpsection orders after each operation in order to determine result regarding the quality of products if excessive spoilage occurs, to find out its causes.

(*viii*) Clean up on jobs, collection of time tickets, blueprints and instruction cards and their return to appropriate section of production control department.

(*ix*) To ensure that the work is forwarded to next department or storeroom, etc.

(*x*) To record the beginning and completion times of jobs on time tickets for calculation of time interval. To forward time ticket to accounts department for preparing wages.

(*xi*) To record and report idle time of men and machines and request for corrective action required.

7.9.2. Dispatch Procedure

The product is broken into different components and components into operations. A route sheet for the part (component) C having three operations on it is shown in Fig. 7.9. The various steps of dispatch procedure for each operation are listed below, in sequence:

ROUTE SHEET PART C
Material
Operation-1
Operation-2
Operation-3

Fig. 7.9. Route Sheet (Dispatching)

(*a*) *Store Issue Order*. Authorise stores (department) to deliver required raw material.

(*b*) *Tool Order*. Authorise tool store to release the necessary tools. The tools can be collected by the tool room attendant.

(*c*) *Job Order*. Instruct the worker to proceed with the operation.

(*d*) *Time Ticket*. It records the beginning and ending time of the operations and forms the basis for worker's pay.

(*e*) *Inspection Order*. Notify the inspectors to carry out necessary inspection and report the quality of the component.

(*f*) *Move Order*. Authorise the movement of materials and components from one facility (machine) to another for further operations.

In addition, there are certain other dispatch aspects which have to be taken care of:

(*i*) All production information should be available before hand.

(*ii*) Various order cards, and specification drawings should be ready.

(*iii*) Equipments should be ready for use.

(*iv*) Progress of various orders should be properly recorded on the Gantt charts or display boards.

(*v*) All production records should be properly maintained.

7.9.3. Centralized and Decentralized Dispatching

Dispatch function may be centralized or decentralized.

In a *Centralized* Dispatch system, a central dispatching department, orders directly to the work station. It maintains a full record of the characteristics and capacity of each equipment and workload against each machine. The orders are given to the shop supervisor, who runs his machines accordingly. In most of the cases, the supervisor can also give suggestions as regards loading of men and machines under him. A centralized system has the following advantages :

(*a*) A greater degree of overall control can be achieved.

(*b*) Effective co-ordination between different facilities is possible.

(*c*) It has greater flexibility.

(*d*) Because of urgency of orders, changes in schedules can be affected rapidly without upsetting the whole system.

(*e*) Progress of orders can be readily assessed at any time because all the information is available at a central place.

(*f*) There is effective and better utilization of manpower and machinery.

In a *Decentralised* dispatching system, the shop supervisor performs the dispatch functions. he decides the sequences of handling different orders. He dispatches the orders and materials to each equipment and worker, and is required to complete the work within the prescribed duration. In case he suspects delay, with due reasons of the same, he informs the production control department. A decentralised dispatching system has the following *advantages :*

(*a*) Much of the red tape is minimized.

(*b*) Shop supervisor knows best about his shop, therefore, the work can be accomplished by the most appropriate worker and the machine.

(*c*) Elaborate reports and duplication of posting can be avoided.

(*d*) Communication gap is reduced.

(*e*) It is easy to solve day-to-day problems.

(*f*) It keeps the natural urge of a section to be self-sufficient.

The advantages of a centralized system, more or less give an idea about the disadvantages of the decentralized system and *vice versa.*

7.10. PROGRESSING OR FOLLOW-UP

Progressing or expediting or follow-up is the process of checking production activities systematically so that production may be carried out according to plan.

7.10.1. Objectives (Purposes)

(*a*) To help to eliminate particular difficulties which are throwing production off the schedule.

(*b*) To speed up the processing of certain orders.

(*c*) To record the flow of materials and tools and to make adjustment if needed.

(*d*) To compare and record amount of work in process with schedule.

(*e*) To record the amount of faulty work and rejection, issue orders for the production of replacements.

(*f*) To record the machine and manpower idleness and to investigate the reasons for it.

(*g*) To record the breakdown, hold-up or stoppage of production activities due to:

(*i*) lack of instructions or blue-print

(*ii*) lack of input materials and components

(*iii*) non-availability of manpower.

7.10.2. Functions

The main function is, to bring up together all the variables of production activities and thus, to show progress or boost production. It is the duty of the follow-up people to see whether the production is being performed as per the schedule and to provide feedback of the production data.

7.10.3. Progress Reporting Techniques

The progress report may be transmitted by the following commonly used techniques:

(*a*) Re-written or pre-typed papers. These are sent through messengers from one department to another.

(*b*) Teletype equipment. It has a keyboard similar to typewriter. Pressing different keys gives rise to electric signals which are transmitted to receiving stations where the message is recorded.

(*c*) Telephone (land-line), mobiles and inter-communication equipments.

(*d*) Radio and Loudspeakers. These are especially used for outdoor applications to control the movement of materials handling equipments and earth-moving machinery.

(*e*) Closed circuit TV for keeping an eye over the process emitting harmful radiations.

7.10.4. Inspection

Inspection is the art of compairing materials, products or performance with the established standards. Inspection is carried out for conforming that quality is of pre-determined standard. During inspection defective parts are rejected so that only products of good quality should reach to the customers. To reduce rejection; inspection is done at various stages starting from the raw materials to the finished product. In order to produce products of good quality, tools and machines are also inspected.

7.11. ADVANTAGES OF BETTER PLANNING AND CONTROL (EFFECTIVENESS OF PPC)

1. Effective Utilisation of Plant Capacity. Production planning and control ensures effective utilisation of plant capacity. All the resources-men, materials, machines ... are put to maximum use. This results in reduced costs and higher returns to the organisation.

2. Steady Flow of Production. Production planning and control ensures regular and steady flow of production. Since all the machines are put to maximum use, there is regular production. This helps to meet regular supplies to customers.

3. Estimate of Resources. It helps to estimate the required resources like men, materials, etc. The estimate in proper quantity and quality is made on the basis of sales forecast. Accordingly production is planned to meet sales requirements.

4. Ensures Optimum Inventory. It prevents overstocking as well as understocking. Necessary stocks are maintained. Stock of raw material is maintained at proper level in order to meet production demands. And at the same time proper stock of finished goods is maintained to meet regular demands from customers.

5. Co-ordination. Proper production planning facilitates co-ordination of activities of various departments. For instance, the marketing department co-ordinates with production department to sell the goods produced because it can get ready supply of quantity and quality of goods required by customers. This results in profitable operations to the organisation.

6. Minimising Wastages. Because of proper inventory of raw materials, there is proper material handling. This helps in minimising wastages of raw material. Also because of effective control, there is production of quality goods. This results in minimum rejects. Thus, proper production planning and control results in minimum wastage. Inventory if held too long can have disasterous effect which is illustrated through the ad of G.E. Information Services on next page.

7. Improvement in Labour Productivity. There is maximum utilisation of manpower. If required, necessary training is provided to the workforce. Benefits of profitable operations are passed on to the workers in the form of increased wages and other incentives. Workers are motivated to perform their best. This results in improved labour efficiency.

8. Adhering to Delivery Schedules. It facilitates adherence to prompt delivery schedules. This is because of regular and steady flow of quality production. Thus, the company can face market competition effectively.

9. Better Work Environment. Proper production planning and control provides better work environment to the workers. Workers get improved working conditions, proper working hours, leave and hoildays, increased wages and other incentives. This is because the company is operating at higher efficiency.

10. Facilitates Quality Improvement. It facilitates quality improvement as necesssary checks are undertaken at regular intervals. Quality consciousness is developed among the employees through training sessions, suggestion schemes, quality circles, etc.

Thus, the overall objective of production planning and control is to improve the efficiency and profitability of the organisation.

7.12. CONCEPTS IN PRODUCT LIFE-CYCLE

The life-cycle concept or 'womb to tomb' concept draws analogy from living organisms. It assumes that every system (product) has a definite life-cycle and it passes through growth, maturity saturation and decline phases. Figure 7.8 shows a typical life-cycle of a product. Similar pattern could exist for the entire production systems. Life-cycle concept enables us to understand various decisions and their interdependence in a better perspective. For example, if some of the strategic decisions like product selection or plant location, which are made at the early stages of systems life-cycle are wrong, then these would continue to influence day-to-day operations planning and control decisions adversely and no amount of day-to-day effectiveness will be able to undo the damage done by poor decisions at initial stages of life-cycle.

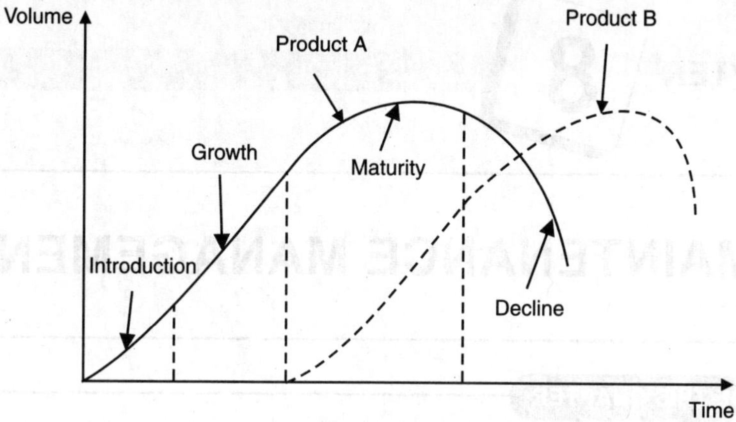

Fig. 7.10. Product Life Cycle

This concept also enables us to be alert to the external environment and start phasing out a dying product and substitute it with a new product well in time so that continued survival of the organisation can be planned. Figure 7.8 shows how introducing new product well in time can cause long-term survival of the organisation even if individual products follow life-cycle pattern. Life span of a product may vary from new months (such as fashion goods) to few decades.

Note: For Production control in job order, batch type and continuous type of production and difference between these control systems—refer to Chapter 1 (Section 1.7).

MODEL QUESTIONS

1. What do you understand by production planning ? Is it a part of PPC ? Explain briefly.
2. Write down and explain the various functions of production planning and control.
3. Define "Process Planning" and describe the various steps involved in process planning.
4. Define routing and write down the functions of routing.
5. Define "Loading" in PPC and explain how a load chart can be drawn.
6. What are the various techniques of scheduling ? Describe each of these techniques in short giving appropriate examples.
7. Define dispatching and write down the various functions of it alongwith the dispatch procedure.
8. "Progressing or follow-up is one of the most important components of production control." Do you agree ? Discuss the involvement of the follow-up technique in production control.
9. Inspection is the last function of PPC. Discuss the process of inspection in the fulfillment of the objectives of production control.
10. Describe machine interference with suitable examples.
11. Explain the concept of Product Life Cycle with suitable examples.

❑❑❑

CHAPTER 8

MAINTENANCE MANAGEMENT

INSIDE THIS CHAPTER

❒ Introduction (Meaning and Definition); ❒ Losses Due to Poor Maintenance; ❒ Objectives of Maintenance; ❒ Types of Maintenance; ❒ Maintenance Organisation; ❒ Activities Involved in Maintenance Management

8.1. INTRODUCTION (MEANING AND DEFINITION)

We all are aware of the term 'maintenance'. Maintenance means protection and upkeep. Any equipment, machine, plant, building, etc. are supposed to produce goods and services of proper quality and in proper quantity. At initial stage of plant and machinery, these produce quality and quantity goods and services. But after use for a certain period, the performance of plant and machinery gradually deteriorates mainly due to *wear and tear, strain, stress and fatigue*. All the plant and machinery need proper protection against such deterioration so that the performance of those can be maintained at high level. The deterioration in performance also leads to overuse of machinery and subsequent loss of production, occurrence of accident and various other losses. In this process, the life of machinery also reduces. The other basic objective of maintenance is to increase the life of the equipment. The capital is scarce resource and hence, it is expected that the capital goods like plant and machinery are maintained properly to ensure effective and longer operational life. Besides, the natural condition like tropical climate, extremely high and low humidity in our country calls for special care of machinery. Unfortunately the actual situation is far below expectation. Because of inadequate maintenance, industries is our country suffers from an unduly high rate of depletion of capital assets and a chronic waste of production capacity which even economically stronger countries could hardly afford. It has been observed that we do not concentrate on maintenance or do not really feel importance of it till such time machine ultimately breaks down. By and large we do not deliberately stop production and carry out maintenance for longer life and better performance. Cost of maintenance is never considered as investment for betterment. On the other hand, it is considered as unnecessary expenditure. Attempt is to avoid or postpone the maintenance expenditure, if the production is on. The lack of proper maintenance practices have serious consequences on the economic development of the country.

Definition (Maintenance)

Maintenance is defined as the activities required to keep a facility in as-built condition continuing to have its original productive capacity by rectifying the damaged and deteriorated parts of the facility to the functional level.

8.2. LOSSES DUE TO POOR MAINTENANCE

In the absence of proper repairs and maintenance, the following types of losses may arise :

1. Lower Productivity. Deterioration of plant and equipment results in frequent interruptions or breakdown which in turn leads to production loss. Such loss is very high in capital-intensive and continuous process industries.

2. Higher Costs. Every interruption in production means cost of idle equipment and wages of idle labour. There is greater wastage of materials and parts. Incompetent and inadequate maintenance result in increase in unit cost of manufacturing.

3. Poor Product Quality. Undetected and uncontrolled wear and tear of machinery results in deterioration of product quality. It is not possible to maintain the standards of workmanship.

4. Destruction of Equipment. When the equipment is allowed to deteriorate beyond repair, the life of machines is reduced. Overuse and careless use of machines cause premature replacement of capital assets. Indian industries suffer from unduly high rate of depletion of capital assets and a chronic waste of production capacity on account of inadequate maintenance. Very often the need for maintenance is considered an unnecessary expense rather than an investment. Lack of proper maintenance practices serves as a constraints in the economic development of the country.

5. Induced Loss. Production loss in one plant leads to negative consequences for both the customers and suppliers alike. When a plant supplies poor quality intermediate goods, additional processing is required or the quality of the final product deteriorates. Frequent breakdowns result is irregular deliveries. In order to ensure uninterrupted production in the face of unstable supplies, the firm has to keep high stock of intermediate components and spare parts. This is an unnecessary locking up of scrace capital and waste of resources.

6. Disguised Loss. When an ill-kept machine breaks down, the plant switches over to an idle equipment to avoid disruption in production. In an industry where installed capacity is underutilized, this may give the impression that excess capacity is being utilized. But in reality there is distinguished poor maintenance. Similarly, the poor quality products may be easily sold if there is a serious shortage of supply. In a seller's market, consumers have to accept whatever quality is available at whatever time. The deterioration in quality and delays in delivery are disguised.

7. Poor Customer Relations. When the quality of product is poor and delivery schedule is not maintained, relations with customers will take a nose dive. This will lead to a permanent fall in sales turnover and loss of market.

8. Poor Staff Morale. Employees lose enthusiasm and interest in work, when the machine or equipment frequently breaks down.

The losses and waste caused by poor maintenance impair economic growth in developing countries like India, where the resources are limited and industrial sector is comparatively small.

8.3. OBJECTIVES OF MAINTENANCE

The main objectives of plant maintenance are as follows :

(*i*) To safeguard investments by minimising the rate of wear and tear.

(*ii*) To preserve and prolong the working life of plant and machinery.

(*iii*) To keep all the productive assets in efficient working condition.

(*iv*) To minimise idle life and loss of production due to breakdown.

(*v*) To maximise efficiency and economy in production through optimum use of all facilities.

(*vi*) To minimise industrial accident through regular inspection and repair of safety devices, etc.

(*vii*) To minimise maintenance and upkeep costs.

8.3.1. Importance of Maintenance

Plant and equipment are heart of a production system. Therefore, proper maintenance of plant and equipment is essential for keeping them in good working condition. This ensures optimum utilization of production facilities and resources. All production equipment is designed to give a predetermined quantity and quality of output. This objective can be realized only when the equipment is supported by proper maintenance. In the absence of timely upkeep, even the best machine will ultimately breakdown untimely. Regular use over a period of time causes wear and tear. Gears become loose, bolts tear off, valves get choked, parts crack, pipes leak, lubricant dries, etc. Therefore, regular inspection, lubrication, repair, etc., must be undertaken to maintain the reliability and efficiency of equipment.

Proper maintenance helps to maximise production performance by preventing breakdowns and by minimising the loss caused by breakdown. In a developing country like ours, maintenance has special significance. The capital is a scarce resource and its best possible use in essential. Moreover, the natural condition like tropical climate and extreme humidity require special upkeep of equipment. Proper maintenance results is better performance and longer life of machinery. It helps to improve reliability of plant and machinery. Maintenance is essential in all manufacturing concerns because machines break down, parts wear out and tools deteriorate. Remember that a stitch in time (*i.e.*, proper maintenence timely) saves *nine hundred* in industry, (original proverb — a stitch in time saves nine).

8.3.2. Failure Analysis

Let us now consider how failures take place. Failure analysis plays a vital role in taking decisions pertaining to maintenance planning and control for effective management subject to the budgetary constrains for such an activity. It is important to identify the nature and occurrence of failures with respect to time. This will be highly significant in designing and ensuring adequate reliable performance. It is seen that the failure rates pattern can be depicted as a bath tub curve as shown in Fig. 8.1.

It is the usual experience with equipments that the failure rate is quite high when the equipment is new or newly installed. The failure rate is greater during the initial starting period of infancy but after this initial phase is over, the failures are relatively quite low. Such a behaviour can be approximated to a 'hyper exponential distribution' as shown in Fig. 8.1. Such behaviour is a sign of the design defects or installation defects. Therefore, those that had these inherent defects failed when the equipments were run. Those that failed much later were those that did not have the design or installation defects. This is somewhat similar to the infant mortality in humans.

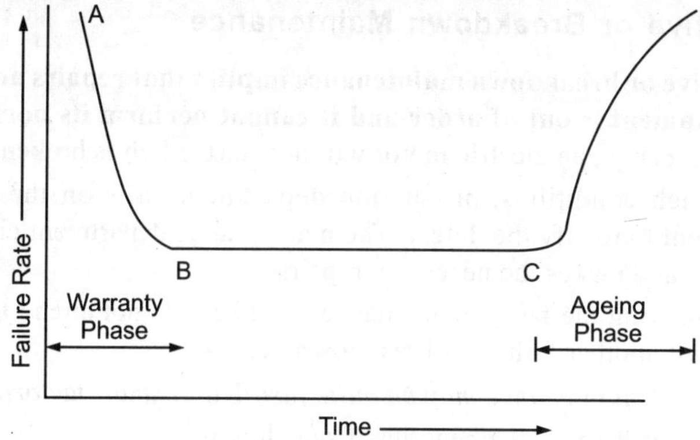

Fig. 8.1. Bath Tub Curve

At the other extreme, many equipments fail due to 'ageing' and wear out. The failure or death may be at a 'mean' or 'average' age, though some should could fail earlier and some later. Such a failure pattern because of wearing out could be represented by a symmetrical bell shaped normal distribution.

We have seen the 'infancy' and the 'old age' phenomenon. In between these two extremes also equipments may fail, but it is neither due to inherent design or installation defects nor due to being worn out. The cause is external to the equipment and therefore, probability of failing is constant and independent of the running time. This phenomenon can be approximated to a 'negative exponential' distribution. This is again similar to the behaviour in humans, where they may die due to external causes such as an epidemic or traffic-accidents while they are neither old nor infants.

The three distributions mentioned above, namely, hyper-exponential, negative exponential and normal, can be confirmed into one distribution, termed as the 'weibull' distribution (named after Weibull who developed it). Hence, the bath tub curve depicting the failure rate can be represented by a weibull distribution. Weibull distributions calculations are very lengthy and tedious. Fortunately weibull graphs are available (though not very common) for determining mean time between failure (MTBF). This would provide good data for determining system reliabilities, availabilities, expected lives, etc. Failure statistics can also be used in the diagnosis of the nature of a recurrent equipment failure and also in the prescription of solutions to maintenance problems.

The prime objective of maintenance is to increase the availability and reliability of a piece of equipment. The availability (A) of a plant can be defined as,

$$A = \frac{T_{up}}{T_{up} + T_{down}}$$

where, T_{up} = the cumulative time of operation in the normal working state

and T_{down} = the cumulative downtime.

8.4. TYPES OF MAINTENANCE

Maintenance may be classified into following categories :

(*i*) Corrective or breakdown maintenance,

(*ii*) Scheduled maintenance,

(*iii*) Preventive maintenance (Planned Maintenance),

(*iv*) Predictive maintenance (Planned Maintenance).

8.4.1. Corrective or Breakdown Maintenance

(*i*) **Corrective or breakdown maintenance implies that repairs are made after the equipment is out of order and it cannot perform its normal function any longer,** e.g., an electric motor will not start, a belt is broken, etc.

(*ii*) Under such conditions, production department calls on the maintenance department to rectify the defect. The maintenance department checks into the difficulty and makes the necessary repairs.

(*iii*) After removing the fault, maintenance engineers do not attend the equipment again until another failure or breakdown occurs.

(*iv*) *This type of maintenance may be quite justified in small factories which* :

- are indifferent to the benefits of scheduling :
- do not feel a financial justification for scheduling techniques ; and
- get seldom (temporary or permanent) demand in excess of normal operating capacity.

(*v*) In many factories *make-and-mend* is the rule rather than the exception.

(*vi*) Breakdown maintenance practice is economical for those (non-critical) equipments whose down-time and repair costs are less this way than with any other type of maintenance.

(*vii*) Breakdown type of maintenance involves little administrative work, few records and a comparative small staff.

There is no planned interference with production programmes.

8.4.1.1. Typical Causes of Equipment Breakdown

(*i*) Failure to replace worn out parts.

(*ii*) Lack of lubrication.

(*iii*) Neglected cooling system.

(*iv*) Indifference towards minor faults.

(*v*) External factors (such as too low or too high line voltage, wrong fuel, etc.)

(*vi*) Indifference towards-equipment vibrations, unusual sounds coming out of the rotating machinery, equipment getting too much heated up, etc.

8.4.1.2. Disadvantages of Breakdown Maintenance

(*i*) Breakdowns generally occur at inopportunate times. This leads to poor, hurried maintenance and excessive delays in production.

(*ii*) Reduction of output.

(*iii*) Faster plant deterioration.

(*iv*) Increased chances of accidents and less safety to both workers and machines.

(*v*) More spoilt material.

(*vi*) Direct loss of profit.

(*vii*) Breakdown maintenance practice cannot be employed for those plant items which are regulated by statutory provisions, for example, cranes, lifts, hoists and pressure vessels.

8.4.2. Scheduled Maintenance

(*i*) Scheduled maintenance is a stich-in-time procedure aimed at avoiding breakdowns.

(*ii*) Breakdowns can be dangerous to life and as far as possible should be minimized.

(*iii*) **Scheduled maintenance practice incorporates (in it), inspection, lubrication, repair and overhaul of certain equipments which if neglected can result in breakdown.**

Inspection, lubrication, servicing, etc., of these equipments are included in the predetermined schedule.

(*iv*) Scheduled maintenance practice is generally followed for overhauling of machines, cleaning of water and other tanks, white-washing of buildings, etc.

8.4.3. Preventive Maintenance (PM)

Introduction

(*i*) **A system of scheduled, planned or preventive maintenance tries to minimize the problems of breakdown maintenance.**

(*ii*) **It is a stich-in-time procedure.**

(*iii*) **It locates weak spots (such as bearing surfaces, parts under excessive vibrations, etc.) in all equipments, provides them regular inspection and minor repairs thereby reducing the danger of unanticipated breakdown. The underlying principle of preventive maintenance is that *prevention is better than cure.***

8.4.3.1. Preventive Maintenance (or PM) Involves

(*i*) Periodic inspection of equipment and machinery to uncover conditions that lead to production breakdown and harmful depreciation.

(*ii*) Upkeep of plant equipment to correct such conditions while they are still in a minor stage.

- Preventive maintenance is practised to some extent in about 75% of all manufacturing companies, but every preventive maintenance programme is tailored as per the requirements of each company.
- The key to all good preventive maintenance programmes, however, is inspection.
- Help can be taken of suitable statistical techniques in order to find how often to inspect.

8.4.3.2. Objectives of PM

(*i*) To minimise the possibility of unanticipated production interruptions or major breakdown by locating or uncovering any condition which may lead to it.

(*ii*) To make plant equipment and machinery always available and ready for use.

(*iii*) To maintain the value of equipment and machinery by periodic inspections, repairs, overhauls, etc.

(*iv*) To maintain the optimum productive efficiency of the plant equipment and machinery.

(*v*) To maintain the operational accuracy of the plant equipment.

(*vi*) To reduce the work content of maintenance jobs.

(*vii*) To achieve maximum production at minimum repair cost.

(*viii*) To ensure safety of life and limbs of the workmen.

8.4.3.3. Procedure of Preventive Maintenance

(*i*) There is no ready-made, on-the-shelf, preventive maintenance programme for any industry. It must be tailor-made-measured and cut to fit the requirements of individual industry or plant ; this is because all industries differ in size, age, location, machinery, resources, layout, and construction.

(*ii*) A well-conceived preventive maintenance programme contains the following elements, features or steps:

(*a*) Who should do PM ?

(*b*) Where to start PM?

(*c*) What to inspect in PM?

(*d*) What to inspect for?

(*e*) How often to inspect (frequency)?

(*g*) Preventive maintenance records

(*h*) Storage of spare parts.

(*i*) Control and evaluation of preventive maintenance.

8.4.3.4. Advantages of PM

(*i*) Reduced breakdowns and connected down-time.

(*ii*) Lesser odd-time repairs and reduced overtime to the maintenance work-force.

(*iii*) Greater safety for workers.

(*iv*) Fewer large-scale and repetitive repairs.

(*v*) Low maintenance and repair costs.

(*vi*) Less stand-by or reserve equipment, and spare parts.

(*vii*) Identification of equipments requiring high maintenance costs.

(*viii*) Lower unit cost of manufacture.

(*ix*) Better product quality and fewer product rejects.

(*x*) Increased equipment life.

(*xi*) Better industrial relations because production workers do not face involuntary lay-offs or loss of incentive bonus because of breakdowns.

8.4.4. Predictive Maintenance (or Condition-based Maintenance)

Condition-based Maintenance is carried out in response to a significant deterioration in unit as indicated by a change in a monitored parameter of the unit condition or performance. It is here that one can make use of predictive maintenance by using a technique called SIGNATURE ANALYSIS, which is intended to continually monitor the health of the equipment by recording systematically signals or information derived from the form of mechanical vibrations, noise signals, acoustic and thermal emissions, changes in chemical compositions, smell, pressure, relative displacement and so on. Scientific collection of these informative signals or signatures, diagnosis and detection of the faults, if any, based on the knowledge hitherto acquired in the field, and judging the severity of the faults for decision-making, all put together, is called '**Signature Analysis**'. The technique involves the use of electronic instrumentation specially designed for the purpose of varied capacities, modes of application and design features. Vibration and noise signals are the most versatile parameters in machine condition monitoring techniques. Periodic vibration checks reveal whether troubles are present or impending. Vibration signature analysis reveals which part of the machine is defective and why? Sound or noise analysis is somewhat similar to vibration analysis. A stock pulse meter is used to monitor the condition of roller bearings.

Condition-based maintenance, thus reduces injuries and fatal accidents caused by machinery as the conditions of machinery are indicated well before hand. It enables the plant to be stopped safely when instant shutdown is not permissible. Moreover, it permits advanced planning to reduce the effect of impending breakdowns and be in time to have necessary spare parts available. However, condition monitoring is not always used because it involves high manpower and monitoring costs and, furthermore, it is difficult to monitor some parameters.

8.5. MAINTENANCE ORGANISATION

A sound organisational setup is necessary for effective administration maintenance activities. A tailor-made organisation should be developed keeping in view the volume and nature of the maintenance work. Every individual should be entrusted with a well-defined maintenance assignment for which he is qualified. The head of the maintenance department should be made responsible to the head of the production or works department. The person-in-charge of the maintenance function is generally called plant engineer or maintenance engineer. The maintenance personnel should have sound training and specialisation. Definite lines of authority and responsibility should be established between the maintenance group and other concerned groups. A typical organisation for maintenance work is given below :

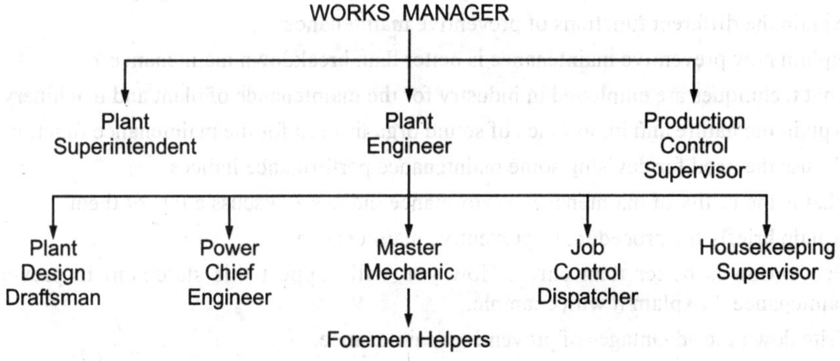

Fig. 8.2. Structure of maintenance organisation

The organisation structure for maintenance may be centralised or decentralised. In centralised setup, all maintenance requests are directed to a central maintenance section which sends suitable team. Under decentralised set up, all major production departments or plants are assigned maintenance crews according to the volume of work and degree of specialisation. In addition, there is a central group responsible for coordination of decentralised groups, stores, machining operations and administration. The choice between the two setups depends on the local conditions and requirements. Maintenance department is primarily a service department.

8.6. ACTIVITIES INVOLVED IN MAINTENANCE MANAGEMENT

The main activities of a maintenance department are as follows :

(*i*) Planning maintenance work on a long-term basis.

(*ii*) Issuing maintenance work orders.

(*iii*) Storing maintenance materials, e.g., tools, spare parts, lubricants, etc.

(*iv*) Deciding inspection methods and routine.

(*v*) Developing and issuing standard instructions.

(*vi*) Measuring efficiency of maintenance.

(*vii*) Controlling maintenance personnel.

(*viii*) Maintaining maintenance records.

MODEL QUESTIONS

1. What do you understand by "Maintenance" ? Should maintenance be regarded just as a repair function ?
2. Describe the importance of maintenance in engineering field.
3. Poor maintenance of plant and machinery leads to various losses in industry. What are these losses ? Discuss briefly.
4. Discuss the objectives of maintenance management in industry. How does maintenance help to increase productivity ?
5. Can you think of some examples from your experience that either conform to or are at variance from the bath tub curve of failure rate phenomena ?
6. Explain the various types of maintenance undertaken in manufacturing operations.
7. Explain the different functions of preventive maintenance.
8. Explain how preventive maintenance is better than breakdown maintenance ?
9. What techniques are employed in industry for the maintenance of plant and machinery ?
10. Explain the nature and importance of sound organisation for the maintenance function.
11. Discuss the need for devising some maintenance performance indices.
12. What is the utility of maintenance performance indices ? Discuss a few of them.
13. Explain briefly the procedure of preventive maintenance.
14. "Prevention is better than cure". How you will support this statement in preventive maintenance ? Explain it with example.
15. Write down the advantages of preventive maintenance.
16. Write short notes on:

    (*a*) Predictive maintenance.

    (*b*) Nature of maintenance problems.

    (*c*) Maintenance organisation.

    (*d*) Tero-technology.

❑❑❑

STATE BOARD EXAMINATION PAPER

Operations Management, 2017

Time : 3 Hrs. *Maximum Marks : 100*

Part-A

Q.1. Attempt any 10 questions. **(10 × 2 = 20)**

(*i*) What is plant layout? (Refer 2.2)

(*ii*) What is product life cycle? (Refer 7.11)

(*iii*) Write importance of forecasting. (Refer 3.4)

(*iv*) What are the error estimates in forecasting? (Refer 3.7)

(*v*) What are the assumption is basic EOQ model? (Refer 4.4)

(*vi*) Discuss the scope of ERP. (Refer 4.11)

(*vii*) Define KANBANs. (Refer 4.9.3.)

(*viii*) Define JIT. (Refer 4.9.2.)

(*ix*) Define meaning of supply chain. (Refer 6.2)

(*x*) Define its meaning of the term project. (Refer 5.1)

(*xi*) Johnson's algorithm is used for.
(Ans. for sequencing of n jobs × 2 machines or 3 machines or more Ref. 7.83)

(*xii*) Define activity. (Refer 5.4)

(*xiii*) What is free float? (Refer 5.7)

(*xiv*) Write down the advantages of sequencing. (Refer 7.8)

Q.2. Attempt any 5 questions. **(5 × 4 = 20)**

(*i*) What are the functions of Production Planning and Control? (Refer 7.3)

(*ii*) Write down the factors affecting plant location decision. (Refer 2.1.1)

(*iii*) Explain the concept of time series in forecasting. (Refer 3.8)

(*iv*) Write meaning and objectives of inventory management. (Refer 4.1 and 4.2)

(*v*) Discuss forecast errors. (Refer 3.7)

(*vi*) What are advantages of sequencing? (Refer 7.8)

(*vii*) What are the functions of purchasing department? (Refer 7.6.3)

(*viii*) What are different types of maintenance system? (Refer 8.4)

Part-B

Attempt any 3 questions. **(3 × 20 = 60)**

Q.3.(*a*) Give brief introduction of material handling machine and its functions. (Refer 2.3.5.)

(*b*) Explain the concept of plant layout and give its classification also. (Refer 2.2 and 2.2.4)

Q.4.(*a*) Describe the difference between short-range an long-range forecast. (Refer 3.5.1, 3.5.3)

(*b*) Explain the ABC inventory and indicate its advantages? (Refer 4.5.2)

Q.5.(*a*) **Distinguish between moving average, exponential smoothing and trend projection methods of forecasting.** (Refer 3.5)

(*b*) **What is e-commerce? How it is helpful in supply chain management?** (Refer 6.10.3)

Q.6.(*a*) **Explain the concept of loading and write down its different types.** (Refer 7.6)

(*b*) **What is critical path and what is its importance in project planning?** (Refer 5.6)

Q.7.(*a*) **How is a master production schedule created and how is it used?** (Refer 7.7)

(*b*) **CPM uses a deterministic time estimate and PERT uses probabilistic time estimate. Explain.** (Refer 5.8)

INDEX

Win Prizes

ATTENTION READERS

Win Prizes

We request for your frank assessment regarding various aspects of the book as given below :

OPERATIONS MANAGEMENT

By: A.P. Verma

Please Fill up the Sheet in Neat and Clean Writing

(i) What topics of your syllabus, important for your examination point of view, are not covered in this book ?

....

....

....

....

(ii) What are the chapters wherein the treatment of the subject matter is not systematic or organised or updated ?

....

....

....

....

(iii) Have you come across misprints/mistakes/factual inaccuracies in the book? Please specify the chapters and the page numbers.

....

....

....

....

(iv) What other books on the same subject have you found/heard better than the present book? Please specify in terms of price and quality.

....

....

....

....

(v) Further suggestions and comments for the improvement of this book.

....

....

....

....

(PS: If need be please attach a separate sheet to write your views)

OTHER DETAILS

(i) Who recommended you this book ? (Please tick (✓) in the box near the option relevant to you.

☐ Teacher ☐ Friends ☐ Bookseller

....

....

....

....

(ii) Name of the recommending teacher, his designation and address :

....

....

....

....

(iii) Name and address of the bookseller you purchased the book from :

....

....

....

....

(iv) Name and address of your institution (Please mention the University or Board as the case may be)

....

....

....

....

(v) Your name and complete postal address :

....

....

....

....

(vi) Course you are enrolled in :

....

....

....

....

The best assessment will be awarded each month: The award will be in the form of our publications as decided by the Editorial Board.

Please mail the filled up coupon at your earliest to :

S.K. Kataria & Sons®
4885/109, Prakash Mahal, Dr. Subhash Bhargav Lane,
Opposite Delhi Medical Association, Daryaganj, New Delhi-110002
Phones: +91-11-23243489, 43551243; Mobile: +91-9871775858
e-mail: katariabook@gmail.com; katariabooks@yahoo.com
Website: www.skkatariaandsons.com